WLANs and WPANs towards 4G Wireless

The Artech House Universal Personal Communications Series

Ramjee Prasad, Series Editor

WLANs and WPANs towards 4G Wireless

Ramjee Prasad
Luis Muñoz

Artech House
Boston • London
www.artechhouse.com

Library of Congress Cataloging-in-Publication Data
A catalog for this book is available from the U.S. Library of Congress.

British Library Cataloguing in Publication Data
Prasad, Ramjee
WLANs and WPANs towards 4G wireless.—(Artech House universal personal
communications library)
1. Mobile communication systems. 2. Wireless LANs.
I. Title. II. Muñoz, Luis.
621.3'8456
ISBN 1-58053-090-7

Cover design by Igor Valdman

International Standard Book Number: 1-58053-090-7
A Library of Congress Catalog Card Number is available from the U.S. Library of Congress.

10 9 8 7 6 5 4 3 2 1

To my wife, Jyoti, to our daughter, Neeli, to our sons, Anand and Rajeev,
and to our granddaughters, Sneha and Ruchika
—Ramjee Prasad

To my wife, Dina, to my mother, Hélène, to my father,
to my mother-in-law, and to the memory of my aunt Rachel
—Luis Muñoz

Contents

Preface

यतो यतो निश्चरति मनश्चञ्चलमस्थिरम् ।
ततस्ततो नियम्यैतदात्मन्येव वशं नयेत् ॥

Yato yato niœcalati
Manaœ cañcalam asthiram
Tatas tato niyamyaitad
    —Âtmany eva vaœeam nayet

From wherever the mind wanders due to its flickering and unsteady nature,
one must certainly withdraw it and bring it back under the control of the self.
    —The Bhagvad Gita (6.26)

This book paves the path toward fourth generation (4G) mobile communication by introducing mobility in heterogeneous IP networks with both *third generation* (3G) and *wireless local area networks* (WLANs), which is seen as one of the central issues in the becoming 4G of telecommunications networks and systems. This book presents a thorough overview of 3G networks and standards and discusses interworking and handover mechanisms between WLANs and the *Universal Mobile Telecommunication System* (UMTS).

    This book is a new, forward-looking resource that explores the present and future trends of WLANs and *wireless personal area networks* (WPANs). This book also provides the discovery path that these infrastructures are following from a perspective of synergies with 3G systems and how they will pave the way for future 4G systems. It is a good resource for learning what performance can be expected from WLANs and WPANs when they support the *Transmission Control Protocol* (TCP)/IP stack. Several critical issues are examined in depth, including IP routing and mobility, the ad hoc concept, IEEE 802.11 and the *high performance WLAN* (HIPERLAN/2) standards, physical (PHY) and *medium access control* (MAC) layers for the main WLAN specifications, the TCP-*User Datagram Protocol* (UDP)/IP protocol stack, and the performance of the TCP-UDP/IP stack over the IEEE 802.11b platform. An entire chapter is devoted to the WPAN domain, where a detailed treatment of Bluetooth and a *second generation* (2G) outlook are provided. Moreover, the book explains how the *performance-enhancing proxy* (PEP) paradigm provides interworking capabilities between WLANs and WPANs and how it enhances performance over these platforms. This practical

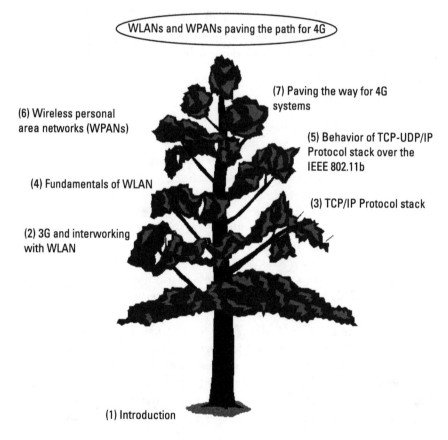

Figure P.1 Illustration of the coverage of the book. The numbers in the branches denote the chapters of the book.

resource is designed to help the researchers save time when planning next generation networks, offering solutions for interworking between WLANs and public cellular networks and for improving the performance of these networks when they support IP. Figure P.1 illustrates the coverage of the book.

This book is intended for everyone in the field of wireless information and multimedia communication systems. It provides different levels of material suitable for managers, researchers, system designers, and graduate students. We hope that all readers will experience the benefits and power of this knowledge.

Acknowledgments

This book presents many of the results obtained throughout the last three years in the area of *wireless local area networks* (WLANs) and under the umbrella of the European project Wireless Internet Networks (WINE). For this reason we would like to thank Ramón Agüero, Johnny Choque, César Espinosa, Marta García, and Luis Sánchez, all of whom belong to the Department of Communications Engineering of the University of Cantabria, for their effort and creative work during this time. Moreover, we would also like to express our gratitude to Verónica Gutiérrez and Francisco Soberón for their help and willing cooperation throughout the development of this book. Finally, within the WINE consortium, we would like to thank colleagues from the companies, laboratories, and universities participating in WINE.

Junko Prasad helped to prepare the manuscript, freeing us from this enormous editorial burden, for which we are immensely grateful to her. Last but not least, we are indebted to Avaya and the IEEE for allowing us to use information concerning their products, as well as material corresponding to publications and standards.

1

Introduction

Marchese Guglielmo Marconi said in 1932, "It is dangerous to put limits on wireless." But even Marconi might not have dreamed what has already been achieved and what may happen next in the field of wireless communications. Looking at these unbelievable, extraordinary, and rapid developments, Ramjee Prasad said in 1999, "It is dangerous to put limits on wireless data rates, considering economic constraints." This rapid development will shrink the world into a *global information multimedia communication village* (GIMCV) by 2020. Figure 1.1 illustrates the basic concept of a GIMCV, which consists of various components at different scales ranging from global to picocellular size.

1.1 Global Information Multimedia Communication Village

A successful operation of the *first generation* (1G) of the wireless mobile communication gave the birth to the concept of the GIMCV. A family tree of the GIMCV system is shown in Figure 1.2 [1–11].

The GIMCV has been evolving since the birth of the 1G analog cellular system. Various standard systems were developed worldwide. Table 1.1 summarizes these analog cellular communication systems.

In the United States, an analog cellular mobile communication service called *advanced mobile phone service* (AMPS) was started in October 1983 in Chicago [12].

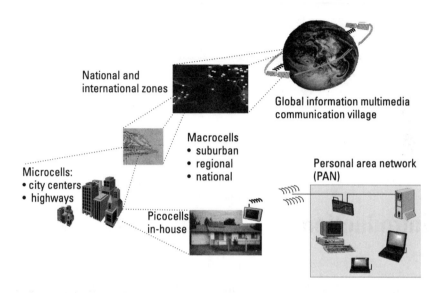

Figure 1.1 Global information multimedia communication village.

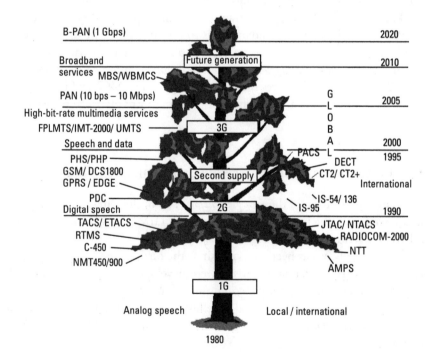

Figure 1.2 Family tree of the GIMCV. Branches and leaves of the GIMCV family tree are not shown in chronological order.

Table 1.1
Summary of Analog Cellular Radio Systems

System	AMPS	NMT-450	NMT-900	TACS	ETACS
Frequency range (mobile Tx/base Tx) (MHz)	824–849/ 869–894	453–457.5/ 463–467.5	890–915/ 463–467.5	890–915/ 935–960	872–905/ 917/950
Channel spacing (kHz)	30	25	12.5*	25	25
Number of channels	832	180	1,999	1,000	1,240
Region	The Americas, Australia, China, Southeast Asia	Europe	Europe, China, India, Africa	United Kingdom	Europe, Africa

System	C-450	RTMS	Radiocom-2000	JTACS/ NTACS	NTT
Frequency range (mobile Tx/base Tx) (MHz)	450–455.74/ 460–465.74	450–455/ 460–465	165.2–168.4/ 169.8–173 192.5–199.5/ 200.5–207.5 215.5–233.5/ 207.5–215.5 414.8–418/ 424.8–428	915–925/860– 870 898–901/843– 846 918.5–922/86 3.5–867	925–940/ 870–855 915–918.5/ 860–863.5 922–925/ 867–870
Channel spacing (kHz)	10*	25	12.5	25/12.5* 25/12.5* 12.5*	25/6.25* 6.25* 6.25*
Number of channels	573	200	256 560 640 256	400/800 120/240 280	600/2,400 560 480
Region	Germany, Portugal	Italy	France	Japan	Japan

*Frequency interleaving using overlapping or *interstitial* channels; the channel spacing is half the nominal channel bandwidth.

In Europe, several cellular mobile communication services were started. In Norway, *Nordic Mobile Telephones* (NMT) succeeded in the development of an analog cellular mobile communication system: NMT-450 [13].

In the United Kingdom, Motorola developed an analog cellular mobile communication system called the *total access communication system* (TACS) based on AMPS in the 1984–1985 period. In 1983, NMT started a modified NMT-450 called NMT-900. C-450, Radio Telephone Mobile System (RTMS), and Radiocom-2000 were introduced in Germany, Italy, and France, respectively.

Meanwhile, in Japan, Nippon Telephone and Telegraph (NTT) developed a cellular mobile communication system in the 800-MHz frequency band and began service in Tokyo in December 1979. Furthermore, a modified TACS that changed the frequency band to adjust for Japanese frequency planning and celled JTACS was also introduced in July 1989. Subsequently, *narrowband TACS* (NTACS), which reduced the required frequency band in half, started service in October 1991.

So far, we described the evolution of the analog cellular mobile communication system. However, the incompatibility of the various systems precluded roaming. This meant that users had to change their mobile terminals when they moved to another country. In addition, analog cellular mobile communication systems were unable to ensure sufficient capacity for the increasing number of users and the speech quality was not good.

To solve these problems, the research and development of cellular mobile communication systems based on the digital radio transmission scheme was initiated. These new mobile communication systems became known as the second generation of mobile communication systems, and the analog cellular era thus is regarded as the first generation of mobile communication systems. Table 1.2 summarizes digital cellular radio systems.

In Europe, GSM, a new digital cellular communication system that allowed international roaming and used the 900-MHz frequency band, started service in 1992. In 1994, DCS-1800, a modified GSM that used the 1.8-GHz frequency band, was launched.

The development of GSM further moved to GSM phase 2+. The most important standardized GSM phase 2+ work items from the radio access system point of view have been [6]:

- *Enhanced full-rate* (EFR) speech codec;
- *Adaptive multirate* (AMR) codec;
- 14.4-Kbps data service;
- *High-speed circuit-switched data* (HSCSD);
- *General packet radio service* (GPRS);
- *Enhanced data rates for global evolution* (EDGE).

Table 1.2

Summary of Digital Cellular Radio Systems

Systems	*Global System for Mobile Communication* (GSM) *Digital Communications System* (DCS)-1800	IS-54	IS-95	*Personal Digital Cellular* (PDC)
Frequency range (base Rx/Tx, MHz)	GSM: Tx: 935–960; Rx: 890–915 DCS-1800: Tx: 1,805–1,880; Rx: 1,710–1,785	Tx: 869–894; Rx: 824–849	Tx: 869–894 Rx: 824–849	Tx: 810–826; Rx: 940–956; Tx: 1,429–1,453; Rx: 1,477–1,501
Channel spacing (kHz)	200	30	1,250	25
Number of channels	GSM: 124 DCS-1,800: 375	832	20	1,600
Number of users per channel	GSM: 8 DCS-1,800: 8/16	3	63	3
Multiple access	TDMA/*frequency division multiple access* (FDMA)	TDMA/FDMA	*Code division multiple access* (CDMA)/FDMA	TDMA/FDMA
Duplex	*Frequency division duplex* (FDD)	FDD	FDD	FDD
Modulation	*Gaussian minimum shift keying* (GMSK)	*π/4 differential quadrature phase shift keying* (DQPSK)	Binary phase shift keying (BPSK)/QPSK	π/4 DQPSK
Speech coding and its rate (Kbps)	*Regular pulse exciting-long term predictive coding* (RPE-LTP) 13	*Vector-sum excited linear predictive coding* (VSELP) 7.95	*Qualcomm code excited linear predictive coding* (QCELP) 8	VSELP 6.7
Channel coding	1/2 Convolutional	1/2 Convolutional	Uplink 1/3 Downlink 1/2 Convolutional	9/17 Convolutional
Region	Europe, China, Australia, Southeast Asia	North America, Indonesia	North America, Australia, Southeast Asia	Japan

Table 1.3 compares the GSM data service.

In North America, the IS-54 digital cellular communication system was standardized in 1989. Subsequently, the standard was revised to include dual-mode services between analog and digital cellular communication systems and reintroduced in 1993 with the title DAMPS, or IS-136. In addition, IS-95, which was the first standardized system based on CDMA, started service in 1993.

In Japan, the digital cellular communication or PDC systems using the 800-GHz and 1.5-GHz frequency bands started service in 1993 and 1994, respectively.

In addition to these digital systems, the development of new digital cordless technologies gave birth to the second-supplement-generation systems, namely, *personal handy-phone systems* (PHSs)—formerly PHPs—in Japan, the *digital enhanced* (formerly European) *cordless telephone* (DECT) in Europe, and *personal access communication services* (PACSs) in North America. Table 1.4 summarizes the second-supplement-generation systems [14, 15] and shows the cordless telecommunications, second generation (CT2) and CT2+. A detailed description of CT2 can be found in [16, 17], where CT2+ is a Canadian enhancement of the CT2 common air interface.

In the *second generation* (2G) of mobile communication systems, the common standardizations of some regions, such as in Europe and North

Table 1.3
Comparison of GSM Data Services

Service Type	Data Unit	Maximum Sustained User Data Rate	Technology	Resources Used
Short message service (SMS)	Single 140 octet packet	9 bps	Simplex circuit	*Standalone dedicated control channels* (SDCCH) or *slow associated control channel* (SACCH)
Circuit switched data	30 octet frames	9,600 bps	Duplex circuit	*Traffic channel* (TCH)
GPRS	1,600 octet frames	171 Kbps	Virtual circuits and packet	Physical data channel (PDCH) (1–8 TCH)
HSCSD	192 octet frames	115 Kbps	Duplex circuits	1–8 TCH
EDGE	—	384 Kbps	Virtual circuits and packet	1–8 TCH

Table 1.4
Summary of Digital Cordless Systems

System	CT2/CT+	DECT	PHS	PACS
Frequency range (base Rx/Tx, MHz)	CT2: 864–868 CT2+: 944–948	1,880–1,900	1,895–1,918	Rx: 1,930–1,990 Tx: 1,850–1,910
Channel spacing (kHz)	100	1,728	300	300
Number of channels	40	10	77	96
Number of users per channel	1	12	4	8
Multiple access	FDMA	TDMA/FDMA	TDMA/FDMA	TDMA/FDMA
Duplex	*Time-division duplex* (TDD)	TDD	TDD	TDD
Modulation	*Gaussian frequency shift keying* (GFSK)	GFSK	$\pi/4$ DQPSK	$\pi/4$ DQPSK
Speech coding	ADPCM 32	ADPCM 32	ADPCM 32	ADPCM 32
Channel coding	None	*Cyclic redundancy code* (CRC)	CRC	Block coding
Region	Europe, Canada, China, Southeast Asia	Europe	Japan, Hong Kong	United States

America, enabled the realization of partial roaming. This feature was a unique point of the 2G systems in comparison with the 1G systems. The advent of a common standard gave users a sense of ease of international roaming. Users have been eager to see worldwide standardization.

During the period 1990–2000, the styles of wired communication as well as wireless communication were both changed by the innovation of digital signal processing. During the period, all information such as voice, data, images, and moving-images could be digitized, and the digitized data could be transmitted through a worldwide computer network such as the Internet. Mobile users were also eager to be able to transmit such digitized data in a mobile communication network. However, in the 2G mobile communication systems, the data transmission speeds are limited, creating the need for new high-speed mobile communication systems. Based on this objective, research and development into *third generation* (3G) mobile communication systems were started in 1995. The research and development that occurred in the 1995–2000 period can be categorized into two areas:

1. International standardized high-speed digital cellular systems with mobility as the second generation;
2. International standardized broadband mobile-access system with low mobility.

In the first area, *international mobile telecommunication* (IMT)-2000 has become the standard. IMT-2000 aims to realize 144 Kbps, 384 Kbps, and 2 Mbps under high mobility, low mobility, and stationary environments, respectively. Figure 1.3 shows an image of the IMT-2000 concept.

In IMT-2000, on the basis of CDMA, three radio-access schemes have been standardized:

1. *Direct sequence CDMA (DSCDMA)-frequency division duplex (FDD)*—this is known as DSCDMA-FDD;

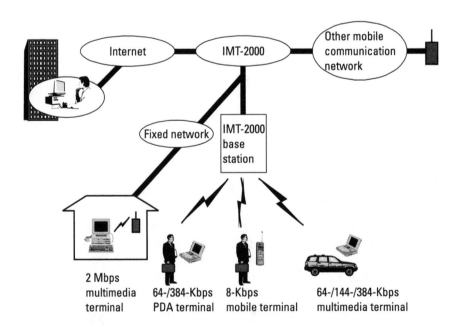

Figure 1.3 Image of IMT-2000.

2. *Multicarrier CDMA* (MCCDMA)-FDD—this is known as MCCDMA-FDD;

3. DSCDMA-TDD.

Wideband code division multiple access (WCDMA) by NTT Docomo and Ericsson and CDMA2000 by Qualcomm were submitted to the ITU [1, 3]. Their basic requirements are shown in Table 1.5. IMT-2000 adopted a CDMA-based system that brought about the capability of offering worldwide roaming by fixing the code transmission rate (chip rate). Moreover, because the data transmission rates of 3G mobile communication systems (144 Kbps–2 Mbps) are much higher than those of the 2G systems (less than 64 Kbps), users can realize moving image–based communication as well as voice and data communication using a mobile terminal.

Several high-speed wireless access systems have been standardized [4]. These basic requirements are shown in Table 1.6. Figure 1.4 shows an image of a high-speed wireless access system. As stated in Table 1.6, most standardized systems can realize transmissions of more than 10 Mbps. It is especially so in the 5-GHz frequency band: an *orthogonal frequency-division multiplexing* (OFDM)–based high-speed wireless access system can realize several tens of megabits per second transmission rates [4]. By using such a mobile access scheme, broadband data transmission rates, such as several tens of megabits per second, can be realized in a wireless communication network as well as a wired network.

Table 1.5
Summary of IMT-2000

	WCDMA	CDMA2000
Frequency	2-GHz band	—
Bandwidth	1.25/5/10/20-MHz (DSCDMA)	1.25/5/10/20-MHz (DSCDMA) 3.75/5-MHz (MCCDMA)
Chip rate	3.84 Mcps (DSCDMA-FDD, DSCDMA-TDD)	3.84 Mcps (DSCDMA-FDD) 3.6864 Mcps (MCCDMA-FDD)
Data rate	144 Kbps (high-mobility environment) 384 Kbps (low-mobility environment) 2 Mbps (stationary environment)	—
Synchronization between base station	Asynchronous/synchronous	Synchronous
Exchange	GSM-MAP based	ANSI-41 based

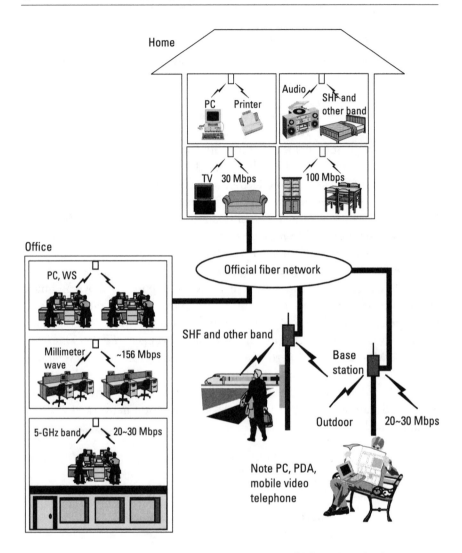

Figure 1.4 Image of high-speed wireless access system. (WS: work station.)

New research and development targets ultra-high-speed wireless access systems that can support data-transmission rates of several tens of megabits per second to hundreds of megabits per second.

Within the European *Advanced Communication Technologies and Services* (ACTS) program, there were four European Union–funded research and development projects ongoing, namely The Magic Wand, a *wireless ATM*

Table 1.6
Summary of Broadband Mobile Access Systems

	Institute of Electrical and Electronics Engineers (IEEE) 802.11 2 GHz	IEEE 802.11 5 GHz	High Performance Local Area Network (HIPERLAN)-2	Multimedia Mobile Project Access Communication Systems Promotion Council (MMAC)
Frequency	2.40–2.4835 GHz	5.150–5.350 GHz 5.725–5.825 GHz	5.150–5.350 GHz 5.470–5.725 GHz	5.150–5.25 GHz
Modulation scheme	*Direct sequence spread spectrum* (DSSS)— DBPSK/DQPSK/*complimentary code keying* (CCK)	OFDM—BPSK/QPSK/16 QAM/64 QAM		
Channel access	CSMA/CA	CSMA/CA	Scheduled TDMA	*Dynamic slot assignment* (DSA)
Duplexing	TDD	TDD	TDD	TDD
Data rate	1,2 Mbps, DBSK, DQSK, 5,5 11 Mbps (CCK)	6, 9 Mbps BPSK; 12, 18 Mbps QPSK; 24, 36 Mbps 16 quadrature amplitude modulation (QAM); 54 Mbps 64 QAM	6, 9 Mbps BPSK; 12, 18 Mbps QPSK; 27, 36 Mbps, 16 QAM; 54 Mbps 64 QAM	6, 9 Mbps BPSK; 12, 18 Mbps QPSK; 27, 36 Mbps 16 QAM; 54 Mbps 64 QAM
Organization	IEEE		*European Telecommunications Standards Institute* (ETSI), *broadband radio access networks* (BRAN)	*Association of Radio Industries and Businesses* (ARIB), MMAC

(WATM) network demonstrator; the *ATM wireless access communication system* (AWACS); the *system for advanced mobile broadband applications* (SAMBA); and wireless broadband *customer premises local area network* (CPN/LAN) for professional and residential *multimedia applications* (MEDIAN) [4, 18–26].

In the United States, a *seamless wireless network* (SWAN) and a *broadband adaptive homing ATM architecture* (BAHAMA), along with two major projects at Bell Laboratories and the *WATM network* (WATMnet), are being developed in the *computer and communication* (C&C) research laboratories of *Nippon Electric Company* (NEC) [18–22].

In Japan, the Communications Research Laboratory (CRL), in the Ministry of Posts and Telecommunications is busy with several research and development projects, such as a broadband mobile communication system [27] in the *super-high-frequency* (SHF) band (from 3 to 10 GHz) with a channel bit rate of up to 10 Mbps, which achieves 5-Mbps transmission in a high-mobility environment where the vehicle speed is 80 km/hr [28, 29]. Moreover, an indoor high-speed wireless LAN in the millimeter-wave band with a target bit rate of up to 155 Mbps [30, 31] has also been researched, and a point-to-multipoint wireless LAN that can achieve a transmission rate of 156 Mbps by using an original protocol named *reservation-based slotted idle signal multiple access* (RS-ISMA) was developed [32].

As a mobile communication system that requires broadband transmission capability, such as several megabits per second to 10 Mbps, in a high-mobility environment, the *intelligent transport system* (ITS) is the most representative example [33–37].

In ITS, there are many communication schemes, of which GPS is the most famous application. However, today, the standardization of the *dedicated short-range communication* (DSRC) system has progressed. The DSRC system uses the *industrial, scientific, and medical* (ISM) band (5.725–5.875 GHz) to realize a short-distance (about up to 30m), vehicle-to-roadside communication system. The image, the applications, and the spectrum allocations for DSRC are shown in Figures 1.5, 1.6, and 1.7, respectively [34].

To realize DSRC, *Comité Européen de Normalisation* (CEN) in Europe, the *American Society for Testing and Materials* (ASTM) and the IEEE in North America, and ARIB in Japan organized standardization committees for DSRC. As for the data transmission scheme, *International Telecommunication Union-Radiocommunication* (ITU-R) recommendation M.1453 suggests two methods: active and backscatter [34]. The requirements are shown in Table 1.7 [34]. Based on the recommendation, several applications are being considered. Figure 1.6 shows some examples of the intended applications. Furthermore, a full-mobility and a quasi-mobile communication system are also being considered.

There are many modulation and demodulation schemes, as well as access protocols used in mobile communication, as described earlier in this section. The relationship between the first, second, and third generation mobile communication systems, high-speed and ultra-high-speed wireless-access systems, and ITS is shown in Figure 1.8.

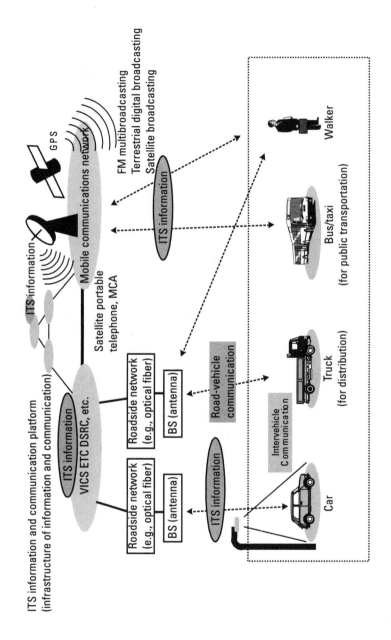

Figure 1.5 Image of DSRC system.

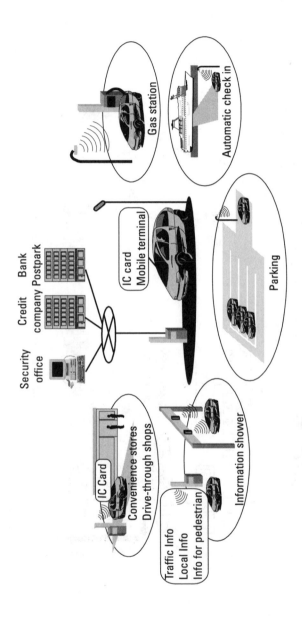

Figure 1.6 Applications image of DSRC system.

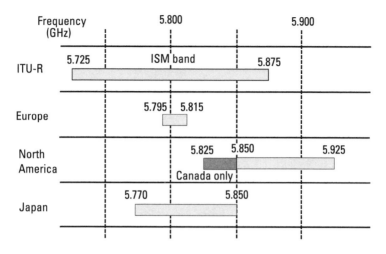

Figure 1.7 Spectrum allocations for DSRC system.

We, therefore, sometimes compare the performance of a new system with that of an old one in a common environment. Computer simulation is one of methods used to evaluate the performance of different systems in a common environment.

1.2 Revenue and Traffic Expectations [38]

The future traffic in terms of transmitted bits will be dominated by data- and packet-oriented traffic [39, 40]. Generally, it is expected that the revenue

Table 1.7
Standardized DSRC System

	Active	**Backscatter**
Organization	ARIB	CEN
RF carrier spacing	10 MHz	1.5, 12 MHz (medium data rate); 10.7 MHz (high data rate)
Allowable occupied bandwidth	Less than 8 MHz	5 MHz (medium data rate); 10 MHz (high data rate)
Modulation method	ASK (uplink/downlink)	ASK (downlink)/PSK (uplink)
Data coding	Manchester code	FMO (downlink)/NRZI (uplink)

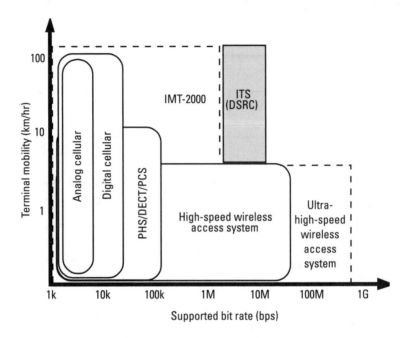

Figure 1.8 Classification of mobile communication systems.

from data services will exceed the revenue from voice in the near future (see Figure 1.9) [41]. Analysts expect that the *average revenue per user* (ARPU) will be shifted from voice to nonvoice services (see Figure 1.10) [42]. This shift from voice to data services and the related revenues will be facilitated by a heterogeneous network architecture to support user needs of *optimally connected anywhere, anytime,* depending on service requirements, user profiles, and location.

The nonreal-time traffic is expected to be mainly asymmetric with the following different types of asymmetry:

- *User-personal view:* The degree of asymmetry for traffic between devices of a *personal area network* (PAN);

- *User-centric view:* The degree of asymmetry for the traffic between a specific user and the network for a specific service;

- *Cell-centric view:* Degree of asymmetry in specific cells;

- *Network view:* Degree of asymmetry in the entire network.

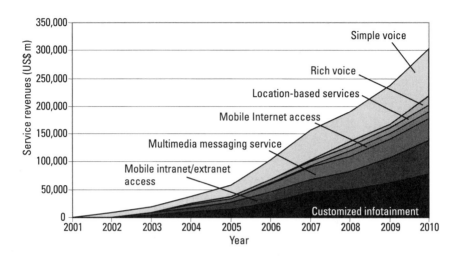

Figure 1.9 Service revenues expectation for 3G for different types of services.

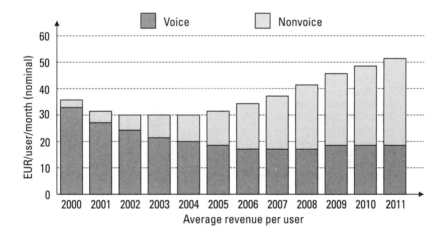

Figure 1.10 Mobile voice and nonvoice ARPU in Western Europe.

The *Universal Mobile Telecommunication Systems* (UMTS) Forum, ITU, and the Japanese Telecommunications Council are expecting increasing traffic asymmetries from the network view according to Figures 1.11 and 1.12 [43, 44]. Figure 1.11 shows the expected spectrum demand for uplink and downlink, whereas Figure 1.12 describes the asymmetry in terms of relative traffic in uplink and downlink. Because the ratio of network-centric

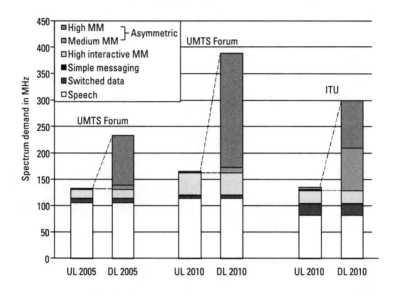

Figure 1.11   Forecast for growth of asymmetric traffic.

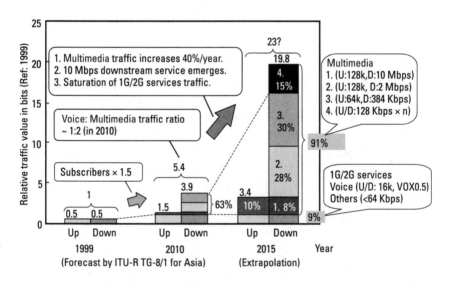

Figure 1.12   Forecast traffic for Region 3 in 2010 and after.

asymmetry is not yet known, future systems have to provide sufficient flexibility to cope with actual differing traffic requirements in uplink and downlink. These requirements have to be taken into account in the development of the new elements of systems beyond 3G [45].

The traffic or system capacity demand is in general inhomogeneously distributed in the deployment area (see Figure 1.13). Therefore, from the network operator's perspective a scalable system architecture is required, which can be adapted to different local traffic and capacity demands with respect to frequency economy and deployment cost. A reasonable solution from economic and service perspectives is a system architecture based on heterogeneous networks, where the different traffic demands are covered by suitable access systems with respect to supported throughput and range.

1.3 Preview of the Book

This book is comprised of seven chapters. The book shows the present and future trends that *wireless local area networks* (WLANs) and *wireless personal area networks* (WPANs) infrastructures will follow with regard to synergies with 3G systems, thus paving the way for the future 4G systems.

In Chapter 2, two important aspects of GIMCV are presented. First, an overview of the 3G networks and standards are introduced and then intraworking and handover mechanisms between the WLAN and UMTS are discussed.

The transport and routing protocols have been presented in Chapter 3. Particular attention is paid to the slow start and congestion avoidance algorithms as well as to the fast retransmit and fast recovery mechanisms. The reason for that is because the performance expected for the Internet stack, in

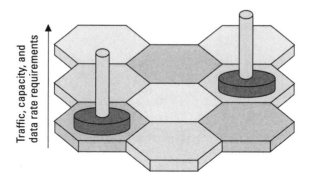

Figure 1.13 Inhomogeneous traffic or system capacity demand in deployment area.

particular for the TCP, running over these wireless infrastructures is heavily dependent on these algorithms.

Also the IP is presented, focusing mainly on aspects like routing and mobility (both micromobility and macromobility). That is, the ad hoc concept and its main challenges, from the perspective of layer three, are presented.

Chapter 4 is devoted to present the main WLANs specifications. In particular the IEEE 802.11 and the HIPERLAN/2 standards are presented. Both the *physical* (PHY) and the *medium access control* (MAC) layers are described for the different specifications.

Chapter 5 shows which performances can be expected on the IEEE 802.11b platforms when the TCP-UDP/IP protocols run over them. In particular, the influence of the wireless errors over the TCP throughput is derived by using a theoretical and practical approach. Also, the UDP protocol behavior over such platform is shown, based on a measurement campaign, and compared with the results obtained by using a theoretical approach.

Chapter 6 presents the works in the domain of the WPAN. Bluetooth is chosen as the 1G WPAN and study in depth. Finally, a 2G WPAN perspective is provided.

In Chapter 7, the *performance enhancing proxy* (PEP) paradigm is proposed as a means to provide internetworking capabilities between WLANs/WPANs and increased performances over such platforms. Our perspective of the future 4G systems based on the collaborative work between WLANs/WPANs with 3G systems is justified.

References

[1] Prasad, R., *CDMA for Wireless Personal Communications,* Norwood, MA: Artech House, 1996.

[2] Prasad, R., *Universal Wireless Personal Communications,* Norwood, MA: Artech House, 1998.

[3] Ojanperä, T., and R. Prasad, (eds.), *Wideband CDMA for Third Generation Mobile Communications,* Norwood, MA: Artech House, 1998.

[4] van Nee, R., and R. Prasad, *OFDM for Wireless Multimedia Communications,* Norwood, MA: Artech House, 1999.

[5] Prasad, R., W. Mohr, and W. Konhäuser, (eds.), *Third-Generation Mobile Communication Systems,* Norwood, MA: Artech House, 2000.

[6] Ojanperä, T., and R. Prasad, (eds.), *WCDMA: Towards IP Mobility and Mobile Internet*, Norwood, MA: Artech House, 2000.

[7] Prasad, R., (ed.), *Towards a Global 3G System: Advanced Mobile Communications in Europe*, Vol. 1, Norwood, MA: Artech House, 2001.

[8] Prasad, R., (ed.), *Towards a Global 3G System: Advanced Mobile Communications in Europe*, Vol. 2, Norwood, MA: Artech House, 2001.

[9] Farserotu, J., and R. Prasad, *IP/ATM Mobile Satellite Networks*, Norwood, MA: Artech House, 2001.

[10] Harada, H., and R. Prasad, *Simulation and Software Radio for Mobile Communications*, Norwood, MA: Artech House, 2002.

[11] Dixit, S., and R. Prasad, (eds.), *Wireless IP and Building the Mobile Internet*, Norwood, MA: Artech House, 2002.

[12] Calhoun, G., *Digital Cellular Radio*, Norwood, MA: Artech House, 1988.

[13] Cox, D. C., "Wireless Network Access for Personal Communication," *IEEE Comm. Mag.*, Vol. 30, December 1992, pp. 96–115.

[14] Padgett, J. E., C. G. Gunther, and T. Hattori, "Overview of Wireless Personal Communications," *IEEE Comm. Mag.*, Vol. 33, January 1995, pp. 28–41.

[15] Kinoshita, K., M. Kuramoto, and N. Nakajima, "Development of a TDMA Digital Cellular System Based on Japanese Standard," *Proc. IEEE VTC'91*, 1991, pp. 642–645.

[16] Tuttlebee, W. H. W., (ed.), *Cordless Telecommunications in Europe*, New York: Springer Verlag, 1990.

[17] Tuttlebee, W. H. W., "Cordless Personal Communications," *IEEE Comm. Mag.*, Vol. 30, December 1992, pp. 42–62.

[18] Prasad, R., "Wireless Broadband Communication Systems," *IEEE Comm. Mag.*, Vol. 35, January 1997, p. 18.

[19] Honcharenko, W., et al., "Broadband Wireless Access," *IEEE Comm. Mag.*, Vol. 35, January 1997, pp. 20–26.

[20] Correia, L. M., and R. Prasad, "An Overview of Wireless Broadband Communications," *IEEE Comm. Mag.*, Vol. 35, January 1997, pp. 28–33.

[21] Morinaga, N., M. Nakagawa, and R. Kohno, "New Concepts and Technologies for Achieving Highly Reliable and High Capacity Multimedia Wireless Communications System," *IEEE Comm. Mag.*, Vol. 35, January 1997, pp. 34–40.

[22] da Silva, J. S., et al., "Mobile and Personal Communications: ACTS and Beyond," *Proc. IEEE PIMRC '97*, September 1997.

[23] Priscoli, F.D., and R. Velt, "Design of Medium Access Control and Logical Link Control Functions for ATM Support in the MEDIAN System," *Proc. ACTS Mobile Comm. Summit '97*, October 1997, pp. 734–744.

[24] Rheinschmitt, R., A. de Haz, and M. Umehinc, "AWACS MAC and LLC Functionality," *Proc. ACTS Mobile Comm. Summit '97*, October 1997, pp. 745–750.

[25] Aldid, J., et al., "Magic into Reality, Building the WAND Modem," *Proc. ACTS Mobile Comm. Summit '97*, October 1997, pp. 734–744.

[26] Mikkonen, J., et al., "Emerging Wireless Broadband Networks," *IEEE Comm. Mag.*, February 1988.

[27] Hase, Y., et al., "R&D Project on Broadband Mobile Communications Using Microwave Band," *Proc. MDMC '96*, July 1996, pp. 158–162.

[28] Harada, H., and M. Fujise, "Experimental Performance Analysis of OCDM Radio Transmission System Based on Cyclic Modified M-Sequences for Future Intelligent Transport Systems," *Proc. IEEE VTC'99 Fall*, Sept. 1999, pp. 764–767.

[29] Harada, H., and M. Fujise, "Field Experiments of a High Mobility and Broadband Mobile Communication System Based on a New Multicode Transmission Scheme for Future Intelligent Transport Systems," *Proc. IEEE VTC'2000 Spring*, May 1999.

[30] Wu, G., et al., "A Wireless ATM Oriented MAC Protocol for High-Speed Wireless LAN," *Proc. IEEE PIMRC '97*, September 1997, pp. 198–203.

[31] Wu, G., Y. Hase, and M. Inoue, "An ATM-Based Indoor Millimeter-Wave Wireless LAN for Multimedia Transmissions," *IEICE Trans. Commun.*, Vol. E83-B, No. 8, August 2000, pp. 1740–1752.

[32] Wu, G., K. Mukumoto, and A. Fukuda, "Analysis of an Integrated Voice and Data Transmission System Using Packet Reservation Multiple Access," *IEEE Trans. Veh. Technol.*, Vol. 43, No. 2, May 1994, pp. 289–297.

[33] Najarian, P. B., "Status Update on ITS Activities in the U.S. and ITS America," *Proc. ITST2000*, October 2000, pp. 7–11.

[34] Ohyama, S., K. Tachikawa, and M. Sato, "DSRC Standards and ETC Systems Development in Japan," *Proc. 7th World Congress on Intelligent Transport Systems*, Torino, Italy, November 2000.

[35] Kim, J. M., "ITS Research and Development Activities at ETRI, Korea," *Proc. ITST2000*, October 2000, pp. 13–18.

[36] Armstrong, L., "New DSRC Standards Under Development for North America," *Proc. 7th World Congress on Intelligent Transport Systems*, Torino, Italy, November 2000.

[37] Rokitansky, C., C. Becker, and Andre Feld, "DSRC Standardisation and Conformance Testing of DSRC/EFC Equipment," *Proc. 7th World Congress on Intelligent Transport Systems*, Torino, Italy, November 2000.

[38] Mohr, W., "Heterogeneous Networks to Support User Needs with Major Challenges for New Wideband Access Systems," *Int. J. Wireless Personal Communication*, Special Issue on Unpredictable Future of Wireless Communication.

[39] Mohr, W., and W. Konhäuser, "Access Network Evolution Beyond Third Generation Mobile Communications," *IEEE Comm. Mag.,* Vol. 38, No. 12, December 2000, pp. 122.

[40] Mohr, W., "Development of Mobile Communications Systems Beyond Third Generation," *Wireless Personal Communications,* Vol. 17, No. 2–3, 2001, pp. 191–207.

[41] Hadden, A., "Great Expectations for 3G," *International Telecommunications,* July 2001, p. 47.

[42] Analysis Research, 2001.

[43] *UMTS/IMT-2000: Assessing Global Requirements for the Next Century,* UMTS Forum, Report No. 6, February 1999.

[44] ITU-R, "Report M [IMT.SPEC]."

[45] ITU-R, "Preliminary draft new Recommendation (PDNR): Vision, framework and overall objectives of the future development of IMT-2000 and of systems beyond IMT-2000," Status: 7th meeting, 2002, Queenstown, New Zealand, February 27–March 5.

2

3G and Its Interworking with WLAN

2.1 Introduction

This chapter presents an overview of the 3G networks and standards as well as interworking and handover mechanisms between the IEEE 802.11 WLAN and UMTS. A detailed description of WLAN is given in Chapter 4.

The *Third Generation Partnership Project* (3GPP) has already specified the Release '99 (R99) standards, which are focused on *asynchronous transfer mode* (ATM) as the backbone network. The recent developments are focusing on all IP-based networks to be standardized for the Release 2000 (R00) of 3GPP. The all-IP network is evolving from packet-switched mobile core network of R99.

With the forecast of over 1 billion mobile users estimated by the end of 2002, packet-based multimedia services, including IP telephony, are expected to account for more than 50% of all wireless traffic. There is a momentum in the industry to evolve the current infrastructure, network services, and the end-user applications toward an end-to-end IP solution capable of supporting *quality of service* (QoS) meeting the needs of the dominant data traffic. At present there are three types of 2G networks: GSM [1, 2], IS-95, and PDC.

There are several 2.5G data transport standards, which are being implemented by many operators. Decisions are based on user demand, spectrum availability, equipment and spectrum license costs, backward compatibility, and assessment of which will be the dominant 3G worldwide standard.

In the future, mobile access to the Internet will be a collection of different wireless services, often with overlapping areas of coverage. No one technology or service can provide ubiquitous coverage, and it will be necessary for a mobile terminal to employ various points of attachment to maintain connectivity to the network at all times. The most attractive solution for such consideration is to utilize high-bandwidth data networks such as IEEE 802.11a/b WLAN whenever they are available and switch to an overlay public network such as UMTS with lower bandwidth when there is no WLAN coverage. Think of a scenario whereby users may wish to be connected to WLAN for low cost and high bandwidth in the home, airport, hotel, or shopping mall but will also want to connect to cellular technologies—for example, GPRS or UMTS—from the same terminal. In particular, the users in this scenario require support for vertical handover (handover between heterogeneous technologies) between WLAN and UMTS.

This chapter describes five possible network-layer-level architectures for interworking and handover between WLAN and UMTS without making any major changes to existing networks and technologies, especially at the lower layers such as MAC and PHY layers [3, 4]. This will ensure that existing networks will continue to function as before without requiring current users to change to the new approach. The implementation involves incorporating new entities like emulators and protocols that operate at the network or higher layers to enable interworking and intertechnology roaming that will be transparent to the mobile user to the extent possible.

Evolution toward 3G is described in Section 2.2. Section 2.3 gives the full details of 3G and its releases (R99 and R00). The 3G-deployment scenarios are discussed in Section 2.4. Section 2.5 presents the 3G impact on the existing network. Section 2.6 describes the general approach to an interconnection philosophy and the essential aspects of making interconnection between IEEE 802.11 WLAN and UMTS. Strategy and consequences behind the five approaches are described in Sections 2.7 through 2.11. Handover issues have been discussed in Sections 2.12 and 2.13. Finally, Section 2.14 presents the conclusions and future directions.

2.2   Evolution from 2G to 3G

There are several 2G-to-3G evolution scenarios for the operators, and some would be content with using 2.5G technologies to make their networks reach

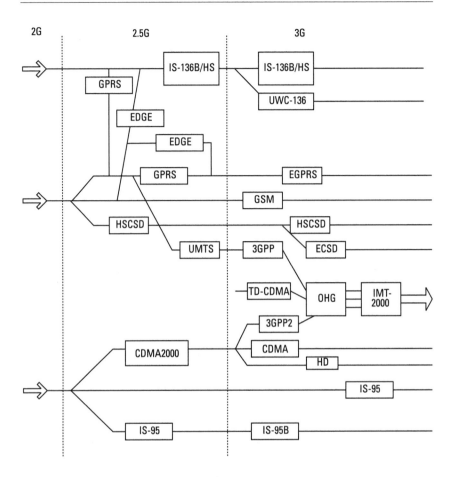

Figure 2.1 Evolution scenarios towards 3G networks.

3G characteristics and features [2, 5–7]. Figure 2.1 shows the different evolution scenarios. Although WCDMA, also known as IMT-2000 or UMTS [8], has emerged as the dominant worldwide standard, other flavors of 3G standards (e.g., CDMA2000 [3, 9]) are still being considered by some operators and countries.

As a solution, the GSM operators are moving towards GPRS data rates of 171 Kbps [2, 8]. TDMA (and some GSM) operators are planning for EDGE [2, 5]—384 Kbps with full mobility. The IS-95 [5] CDMA operators are considering *single carrier radio transmission technology* (1xRTT) [10] (144-Kbps standard). The 1xRTT is interim step towards CDMA2000.

2.3   3G and Its Releases

The 3G will provide mobile multimedia, personal services, the convergence of digitalization, mobility, the Internet, and new technologies based on the global standards [5–7, 11]. The end user will be able to access the mobile Internet at the bandwidth (on demand) from 64 Kbps to about 2 Mbps. From a business perspective, it is the business opportunity of the twenty-first century.

The international standardization activities for 3G is mainly concentrated in the different regions in the *European Telecommunications Standards Institute* (ETSI) *Special Mobile Group* (SMG) in Europe, *Research Institute of Telecommunications Transmission* (RITT) in China, *Association of Radio Industry and Businesses* (ARIB) and *Telecommunication Technology Committee* (TTC) in Japan, *Telecommunications Technologies Association* (TTA) in Korea, and *Telecommunications Industry Association* (TIA) and T1P1 in the United States.

Details of all proposals for IMT-2000 are available in [12]. The international consensus building and harmonization activities between different regions and bodies are currently ongoing. A harmonization would lead to a quasi-world standard, which would allow economic advantages for customers, network operators, and manufacturers. Therefore, two international bodies have been established: 3GPP and 3GPP2.

The 3GPP was established to harmonize and standardize in detail the similar ETSI, ARIB, TTC, TTA, T1 WCDMA, and related TDD proposals [13–15]. The 3GPP decided to base its evolution to an IP core network on GPRS. The GPRS–based approach provides packet data access in 3GPP. The 3G.IP forum initiated the early work on an all IP-network in early 1999, but all the work has since been moved to the 3GPP [13]. The UMTS network architecture is an evolution of the GSM/GPRS. The network consists of three subnetworks: *UMTS terrestrial radio access network* (UTRAN), *circuit-switched* (CS) domain, and *packet-switched* (PS) domain.

The UTRAN consists of a set of *radio network subsystems* (RNSs) connected to the *core network* (CN) through the Iu interface. If the CN is split into separate domains for circuit- and packet-switched core networks, then there is one Iu interface to the circuit-switched CN (Iu-CS) and one Iu interface to the packet-switched CN (Iu-PS) for that RNS, as shown in Figure 2.2.

An RNS consists of a *radio network controller* (RNC) and one or more node Bs. A node B is connected to the RNC through the Iub interface. Inside the UTRAN, the RNCs in the RNSs can be interconnected together through the Iur interface. The Iu and Iur are logical interfaces, which may be provided via any suitable transport network.

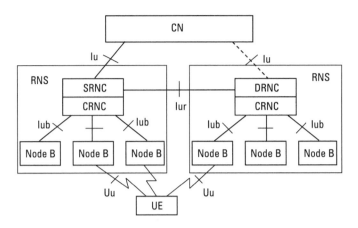

Figure 2.2 UTRAN architecture.

A node B can support one or more radio cells. A node B may support *user equipment* (UE) based on FDD, TDD, or dual-mode operation. During macro diversity (soft handover), a UE may be connected to a number of radio cells of different node Bs or RNSs. Each RNS is responsible for the resources of its set of radio cells and for handover decisions. The controlling part of each RNC (CRNC) is responsible for the control of resources allocated within node Bs connected to that RNC. For each connection between a UE and the UTRAN, one RNC is the serving RNC. When required, drift RNSs support the serving RNC by providing radio resources within radio cells connected to that drift RNC. Combining/splitting for soft handover may be supported within node B, drift RNC, or serving RNC. Softer handover provides better performance but is only possible within node B, between radio cells connected to that node B.

Any RNC can take on the role of serving RNC or drift RNC, on a per-connection basis for a UE. This supports macro diversity (soft handover) when the UE roams into another RNS. Eventually a relocation process (separate from handover) may be used to reroute the Iu connection to the new RNS, after which the drift RNC becomes the serving RNC for the UE. *Radio access bearers* (RABs) are provided between the UE and CN (via the Uu radio interface, UTRAN internal interfaces, and Iu interface) for the transport of user data. Control plane protocols provide the control of these RABs and the connection between the UE and the network. Control plane protocols over Uu would be carried between *radio resource control* (RRC) entities in the UE and UTRAN.

During 2000, 3G was split by 3GPP (see Table 2.1) into two releases: R99 and R2000. R99, also known as Release 3, of the UMTS system supports WCDMA access and ATM-based transport. UMTS R00, split into two releases (Release 4 and Release 5 [13]), defines two *radio access network* (RAN) technologies, a *GPRS/EDGE radio access network* (GERAN) and a wideband CDMA RAN (R3 UTRAN). Both types of RANs connect to the same packet-switched CN (an evolution of the GPRS network) over an Iu interface. One main objective of UMTS R00 is to have the option of all-IP-based CN architecture, thus setting the tone for UMTS standardization in 2000 and beyond. Benefits expected from this approach include the ability to offer seamless services through the use of IP, regardless of means of access, simultaneous multimedia services, and rapid service deployment, in addition to synergy with generic IP developments and reduced cost of service. However, the all-IP architecture in UMTS R00 (Releases 4 and 5) must support the services and capabilities of R99, R00, and beyond. It must ensure an evolution path with sufficient backward compatibility.

The 3GPP2 [9] was also established for the CDMA2000-based proposals from TIA and TTA. Technical specification work for CDMA2000 standardization is being done within 3GPP2 in the following steps:

- CDMA2000 1x, which is an evolution of cdmaOne, supports packet data service up to 144 Kbps.

Table 2.1
3G Releases

3G Release	Abbreviated Name	Specification Version Number	Freeze Date (Indicative Only)
Release 6 (will be TR 21.104)	Rel-6	6.x.y	Scheduled June 2003
Release 5 (TR 21.103)	Rel-5	5.x.y	March 2002
Release 4 (TR 21.102)	Rel-4	4.x.y	March 2001
Release 2000	R00	4.x.y	See table footnote
—		9.x.y	
Release 1999 (TR 21.101)	R99	3.x.y	March 2000
		8.x.y	

Note: The term *Release 2000* was used only temporarily and was eventually replaced by the term *Release 4*.

- CDMA2000 *single carrier evolution-data only* (1xEV-DO) introduces a new air interface and supports high-data-rate service on downlink. It is also known as *high-rate packet data* (HRPD). The specifications were completed in 2001. It requires a separate 1.25-MHz carrier for data only. The 1xEV-DO provides up to 2.4 Mbps on the downlink, but only 153 Kbps on the uplink.

- CDMA2000 *single carrier evolution and voice* (1xEV-DV) will introduce new radio techniques and an all-IP architecture for radio access and CN. The completion of specifications was expected in 2003. It promises data rates up to 3 Mbps.

SK Telecom and LG Telecom from Korea were the first operators to launch CDMA2000 1x in October 2000. Since that time, only a few operators have announced CDMA2000 1x service launches. Some operators recently announced setting up CDMA2000 1xEV-DO trails [10].

The network architecture for a CDMA2000 network is shown in Figure 2.3. The basic architecture is quite similar to the GSM/UMTS architecture. The main differences are in the packet domain, where a *packet data switching node* (PDSN) is used. It has a similar role to the *serving GPRS support node* (SGSN) and *gateway GPRS support node* (GGSN) in UMTS. Mobility management within 3GPP2, however, is based on mobile IP (RFC2002) instead of GPRS mobility management in GSM/UMTS PS networks. Furthermore, *American National Standards Institute* (ANSI)-41 MAP signaling is used instead of GSM *mobile application part* (MAP) signaling. Activities have started in 3GPP2 for evolution toward an all-IP network, similar to the IMS activities in 3GPP.

2.3.1 Release 3 (R3)

Release 3 (or R99) is composed of the UTRAN attached to two separate UMTS CN domains, as shown in Figure 2.4.

1. A circuit-switched domain based on *enhanced GSM mobile switching centers* (E-MSCs) consists of the following network elements:
   - *2G/3G mobile-services switching center* (2G/3G-MSC), including the *visited location register* (VLR) functionality;
   - *Home location register* (HLR) with *authentication center* (AC) functionality.

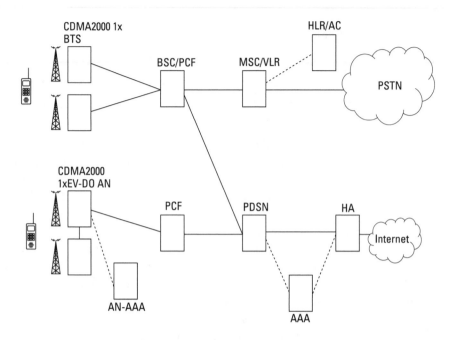

Figure 2.3 CDMA2000 1x and CDMA2000 1xEV-DO network.

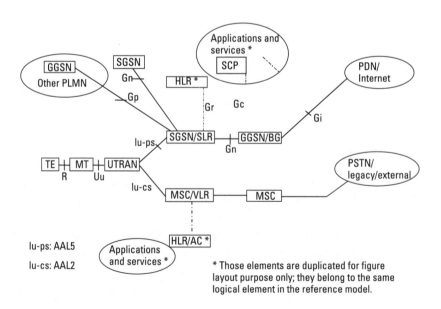

Figure 2.4 Network architecture of UMTS R3.

2. A packet-switched domain built on enhanced GPRS support nodes (E-GSNs) [12, 16] consists of (or involves) the following network elements:

- 2G/3G-SGSN with *subscriber location register* (SLR) functionality;
- *Gateway GPRS supporting node* (GGSN);
- *Border gateway* (BG).

The HLR holds subscriber data and supports mobility in both domains. Two distinct instances of the Iu interface are used between the access and the core network. The hybrid nature of UMTS R3 appears in several aspects. It is most obvious in the transport and call control planes. From an end-to-end connectivity point of view, on one hand UMTS offers switched circuits toward the *public switched telephone network* (PSTN) and *integrated services digital network* (ISDN), mainly to be used for voice communications. On the other hand, IP packet connectivity is provided as a pure network-layer service between a UMTS mobile station and an Internet host. The former is complemented by a sophisticated GSM/UMTS-specific service architecture based on *intelligent network* (IN) principles, mainly for a wide range of supplementary and value-added voice services. By contrast, the latter is confined to cellular radio and mobility-enhanced bearers, although opening the stage for a variety of IP-based applications.

The circuit-switched domain of UMTS R3 builds on the master/slave paradigm [13] of legacy GSM, inherited from the PSTN/ISDN, with the MSC acting as master and the mobile terminal as a slave. The transport plane that physically transports voice is separated from *signaling system #7* (SS7)–based call control or signaling plane that transports signaling messages and ensures advanced features for voice calls.

2.3.2 Release 4 (R4)

The main focuses of R4, with respect to R3, are the following:

- Hybrid architecture—ATM-based UTRAN (currently a workgroup is busy defining IP-based RAN) and IP/ATM-based CN;
- GERAN (support for GSM radio including EDGE);
- Enhanced services provided using toolkits (e.g., CAMEL, MExE, SAT, VHE/OSA);
- Backwards compatibility with R99 services;

- Enhancements in QoS (real-time PS services), security, authentication, and privacy;

- Support for interdomain roaming and service continuity.In R4, the circuit-switched domain is split into a separate signaling plane (MSC server) and transport plane [media gateway (MGW)], which means introduction of new entities. Standardized protocols will be used:

  - Q.BICC or SIP-T (for inter-MSC signaling);

  - H.248/MeGaCo (for MSC server to MGW signaling).

This enables cost reduction by optimizing transport resources. The splitting of MSC into MSC server for signaling and MGW for transport makes R4 scalable, reliable, and cost efficient with respect to R3 (see Figure 2.5). A MSC server is able to support several MGWs. If the serving MGW for a specific connection goes down, the MSC server is in state to reroute the traffic through a different MGW. As an implementation option, it is possible to have a *many-to-many* (m:n) relationship between MSC servers and

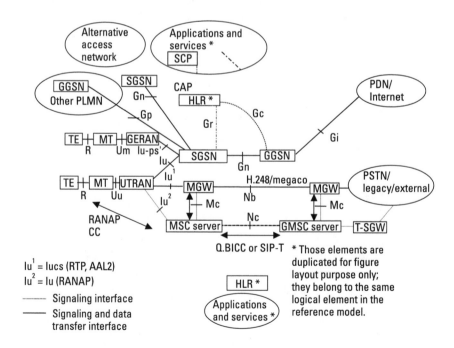

Figure 2.5 Network architecture of UMTS R4.

MGWs. This allows for an efficient allocation of user plane resources as the MSC servers can loadbalance between multiple MGW.

2.3.3 Release 5 (R5)

R5 will mainly focus on the new IP multimedia CN subsystem. IPv6 support is optional for R4 and mandatory for R5. The R5 solution will introduce the all-IP environment, including two major benefits:

- *Transport:* utilization of the IP transport and connectivity with QoS throughout the network;
- *End-user services:* with *session initiation protocol* (SIP), there are possibilities to offer a wide range of totally new services that are not possible to implement in the R4 or earlier releases.

In the end-user service perspective, the implementation of R5 into the network is an add-on. As new terminals are required for SIP services, it is clear that the old GSM and R3 services and subscribers still need to be supported. However, as the functions of classic SS7 call control and the IP call control are so different, it can be foreseen that it is not possible to integrate these functions into the same network element. Thus, a *call state call function* (CSCF), which is basically an SIP server, needs to be introduced.

R5 architecture (see Figures 2.6 and 2.7) is still under discussion. But the general trend is to split the packet-switched domain into control and transport planes.

The benefits of having split SGSN architecture are as follows:

- Flexibility to allocate processing capacity for traffic and for control in different locations;
- Flexibility to independently scale the control plane and the user plane by increasing or decreasing the number of nodes required to handle the corresponding traffic;
- Allows an independent evolution and upgrade of nodes in the user plane and the control plane as the corresponding technology evolves.

As an implementation option, it is possible to have a m:n relationship between SGSN servers and SGSN-GW. This allows for an efficient

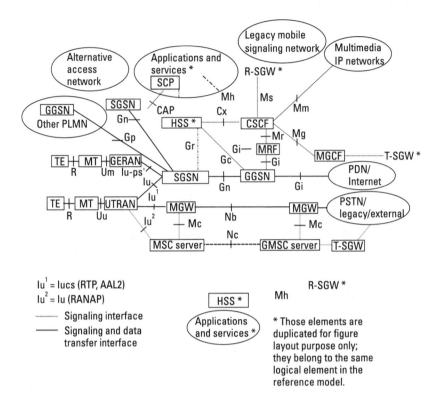

Figure 2.6 Network architecture of UMTS R5.

allocation of user plane resources, as the SGSN servers can set up new *Packet Data Protocol* (PDP) contexts to multiple SGSN-GW for load balancing.

2.4 3G Deployment Scenario

As mobile operators approach the evolution towards 3G, many are examining the continued use of circuit-switched technology within their core networks. Due to the current global economic recession, market uncertainties, and what is today regarded as significant business risks involved with 3G, most operators are trying to reduce and optimize their capital and operational investment in their next generation networks. Even though a lot of money has been invested in the 3G licenses across Europe (i.e., sunk cost), hesitance is seen in quickly investing too much money in what is a relative

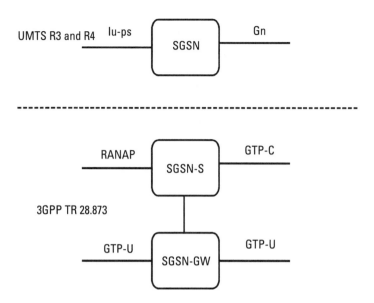

Figure 2.7 SGSN functionality is split into SGSN server for control plane and SGSN-GW for the transport plane.

new technology where the standardization is barely stabilized. It is very interesting to realize that most 3G investments have been made before GPRS has become a commercial success. This is due to the relatively late deployment of commercial volume GPRS handsets/mobile stations and the delayed introduction of mobile packet data into the market. This uncertainty alone puts a tremendous risk in all 3G business plans and should drive the 3G deployment scenario.

The 3G deployment scenarios are driven by two important assumptions:

- There will be a significant shift from voice-centric to data- and multimedia-centric services, which make up the traffic mix within these networks. This is illustrated by the current growth of SMS and expected growth in *Wireless Application Protocol* (WAP) usage in conjunction with GPRS.
- Operators can earn a profit selling 3G services.

Thus, it is clear that current radio access and circuit CN architectures and technologies do not provide an appropriate and efficient infrastructure for the delivery of bursty packet-based data services such as the Internet, *mobile commerce* (m-commerce), and corporate *virtual private networks* (VPNs).

To date, some mobile operators have created portfolios of traditional and next-generation services by building multiple networks. It is common to find a mobile operator using a *time division multiplex* (TDM) network to support voice, an ATM or frame relay network to support GPRS, and an IP network to support new features. Of course, using multiple network infrastructures to support multiple services is costly. In addition, building services that require combining diverse network technologies becomes exceptionally difficult and tedious because they must be manually provisioned.

The use of a single homogenous network based on IP becomes the logical choice for the delivery of seamless services within mobile networks. It allows these services to span the voice, data, and video domains—thus migrating the mobile network towards a true multimedia capability. The migration from circuits to packets is achieved by migrating all of the services and applications within the mobile network onto packet based (IP or IP+ATM) network. This is most logically achieved from the core network with migration outwards towards the RAN.

In the UMTS network architecture, *ATM adaptation layer* (AAL)2-AAL5/*asynchronous transfer mode* (ATM) (R99), or IP (R00) are used. Today it is not easy to determine at what point in time the all-IP model will be introduced. Furthermore, it is likely that R3/4 and R5 will coexist in a large number of mobile operators' networks for quite some time. This is one of the reasons for deploying network architecture based on *multiprotocol label switching* (MPLS), as both IP and ATM are perfectly supported. Figure 2.8 shows a possible network deployment scenario for a 2G mobile operator. With MPLS it is possible to utilize a single physical infrastructure, switches, and transmission links, for both types of traffic: native ATM and IP. Network resource allocation, namely switching capacities and transmission bandwidths, is controlled by a few network management commands. No forklift upgrades will be necessary when gradually moving from an R3 to R5 architecture. This will protect the initial investments of mobile operators and give them total flexibility for the introduction of new services.

2.5 Impact on the Existing Network

In this section, the 3G impact on the existing network architectures will be discussed. It is important to realize that even if most legacy networks are supporting the same access technology, their architectures often are very

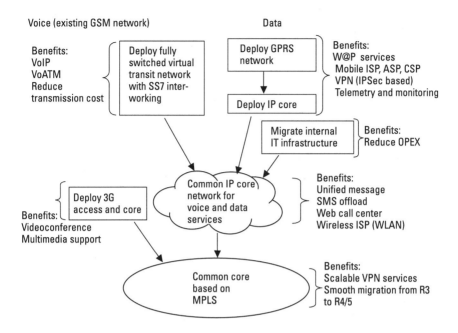

Voice (existing GSM network)　　　　　　　　Data

Benefits:
VoIP
VoATM
Reduce
transmission cost

Deploy fully
switched virtual
transit network
with SS7 inter-
working

Deploy GPRS
network

Deploy IP core

Benefits:
W@P services
Mobile ISP, ASP, CSP
VPN (IPSec based)
Telemetry and monitoring

Migrate internal
IT infrastructure

Benefits:
Reduce OPEX

Deploy 3G
access and core

Common IP core
network for
voice and data
services

Benefits:
Videoconference
Multimedia support

Benefits:
Unified message
SMS offload
Web call center
Wireless ISP (WLAN)

Common core
based on
MPLS

Benefits:
Scalable VPN services
Smooth migration from R3
to R4/5

Figure 2.8　A possible network deployment scenario for 2G mobile operators.

different. Thus, it can be noted that existing mobile networks architectures are very different in nature. They can be characterized by the human experience pool responsible for the architecture, network age and size, vendor or vendors, and whether they interface with fixed networks (either PSTN or IP). The understanding of these facets of the legacy networks (see Figure 2.9) will, together with the need for cost-efficient solutions, be a major driver for 3G deployment and the impact of 3G network rollout on the existing legacy networks. For example, if a mobile network operator already has an ATM backbone network and considerable experience with ATM engineering, it should use this synergy to reduce deployment cost and reduce investments in new transmission and switching equipment. A legacy network that does not have an existing ATM infrastructure could benefit from architecting an IP backbone, although ATM would still need to be supported for uplink to the Iu,CS and Iu,PS. It should also be noted that it is often easier to find good IP engineering skills in comparison with ATM experience. For a single-vendor legacy network, considerable synergy could be found in operational and maintenance experience as well as network monitoring systems by choosing the existing vendor for a 3G network. The initial ease of integration and few

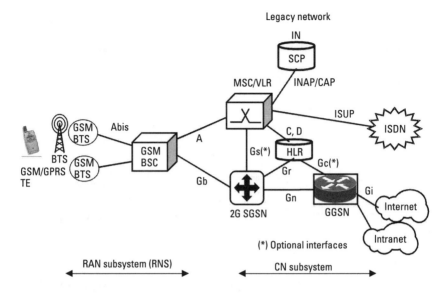

Figure 2.9 2G legacy network.

interoperability issues would also be expected. The big question in this single-vendor scenario is whether the equipment and service pricing will be attractive enough when compared to the addition of a new vendor for the 3G network. The legacy network will be a significant boundary condition for architecting the 3G network. Only Greenfield operations, where no legacy network is present, or an operation with very recent legacy operations might deploy the theoretical architectures (i.e., such as all-IP or near-IP networks with the state-of-the-art IP QoS implementations) presented in standardization bodies.

With all the changes between the GSM network and the 3G network, there are two major impacts on the existing network:

1. *Handover:* It is assumed that the handover decision is always made inside GERAN. For inter-GERAN handover, functions that set up a path within the CN are required. Depending on the handover type, different changes in GSM are required. The backward handover, where the handover signaling is performed through the old base station, is similar to the current GSM handover. For the forward handover, the mobile station initiates the handover through

the new base station. When trying to avoid corner effect, in which the connection to the old station is lost, a very fast handover is required to prevent the blocking of the existing users in another cell. Forward handover requires a large number of changes in GSM.

2. *Transmission infrastructure:* The transmission infrastructure has to meet new requirements imposed by wideband services. Because the data services are bursty and often asymmetric, the transmission solution has to be able to efficiently multiplex different types of information. ATM will provide efficient support for transmission of bursty wideband services. However, because ATM was originally designed for very high-speed transmission in the fixed network, some modifications may be needed to accommodate cellular-specific infrastructure requirements.

There are several possible scenarios for mobile operators to migrate from GSM/GPRS to UMTS. As mentioned earlier in this chapter, the complexities of the 3G migration depends to a high degree on the legacy mobile network and to what extend a single-vendor environment is in place and will remain after the 3G migration. In a multivendor environment, the migration could be considerably more complicated due to mismatches in software feature support and to what 3GPP technical specification release the various vendors adhere to. It is therefore to be expected that in a multivendor environment, the operator will have to compromise the architecture and the services that initially will be launched. Furthermore, where the 2G and 3G vendor differs, interfaces might have to be reconsidered with the possible result of service touch-and-feel changes. A typical example is the interfaces between the HLR and the SGSN (Gr)/GGSN (Gc), and MSC/VLR and SGSN (Gs). Moreover, one vendor's software release package might differ substantially from another vendor's release (after all, with open interfaces features will be what differentiates the various vendors) and could allow for only basic services to be launched or result in significant development work to allow for feature match. A good example of this particular issue is in legacy networks with the open MAP interface between the MSC/VLR and the HLR (i.e., C & D interfaces).

Theoretically, it is possible to interface vendor X HLR with a vendor Y MSC/VLR, but only the basic features could be explored due to feature mismatch. In practice, an operator would always vendor match the MSC/VLR and HLR in order to get the maximum out of the architecture and network infrastructure.

1. To upgrade its existing GSM/GPRS CN for UMTS use, in this case 2G and 3G network share the same core infrastructure, as shown in Figure 2.10. The possible impact on the existing network includes:

   • Redimension of the existing CN to be able to support 3G broadband services;

   • Optimization of the transmission network for a suitable traffic mix;

   • Network management system.

2. To deploy an independent 3G CN from the 2G CN, as shown in Figure 2.11, in this case 2G and 3G network will have minimum impact on each other. In a multivendor scenario, *interoperability tests* (IOTs) will be needed depending on the architecture—for example, Gc (between HLR and GGSN), Iu,PS (RAN and SGSN), and Gr (HLR and SGSN).

2.6　Interworking System Architectures

By connecting the IEEE 802.11 WLAN to a UMTS CN as a complementary radio access network, a second form of mobile packet data services is provided by this heterogeneous IP-based system.

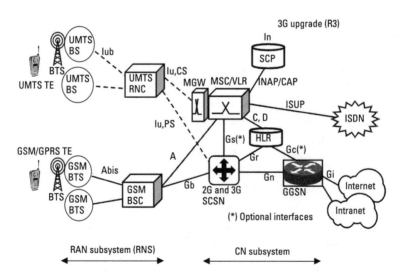

Figure 2.10　Common CN for both 2G and 3G.

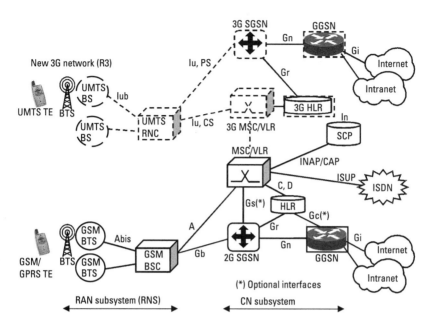

Figure 2.11 Independent 2G and 3G networks.

Figure 2.12 represents the five interconnection points between WLAN and UMTS. These interconnection architectures involve minimum changes to the existing standards and technologies, especially for MAC and PHY layer to ensure that existing standards and networks continue to function as before. The first two interconnections in Figure 2.12 will always have inter-action between WLAN *access point* (AP) and the packet-switched part of the UMTS CN [8]. This means that the gateway to the IEEE 802.11 WLAN network is attached to the packet-switched domain. This interconnection is possible through the *3G serving GPRS supporting node* (3G-SGSN) entity and gateway *GPRS supporting node* (GGSN) entity, which are the elements of the UMTS *packet switched core network* (PS CN). In both cases the WLAN network appears to be a UMTS cell or routing area, respectively. The UMTS network will be a master network and the IEEE 802.11 WLAN network will be a slave network. This means that the mobility management and security will be handled by UMTS network, and the WLAN network will be seen as one of its own cells or routing areas. This may require dual mode *personal computer memory card international association* (PCMCIA) cards to access two different physical layers. In addition, all traffic will first reach the UMTS

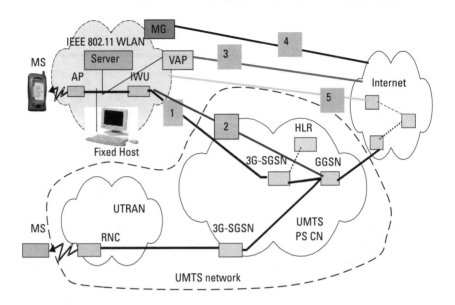

Figure 2.12   Interconnection architectures between IEEE 802.11 WLAN and UMTS.

3G-SGSN or 3G-GGSN before reaching their final destinations, even if the final destination were to be in the WLAN home network.

In the third interconnection, the *virtual access point* (VAP) reverses the roles played by the UMTS and WLAN in the first two architectures. This is called a *tight coupling* because there is always interaction between both networks. Here, the IEEE 802.11 WLAN is a master network, and the UMTS is the slave network. Mobility management is according to the WLAN and *Interaccess Point Protocol* (IAPP) is the protocol that is specified for this management.

In the fourth interconnection architecture, a mobility gateway/mobile proxy (MG) is employed in between the UMTS and IEEE 802.11 WLAN networks. These are both peer-to-peer networks. The MG is a proxy that is implemented on either the UMTS or the WLAN sides and will handle the mobility and routing.

The fifth interconnection architecture is based on mobile IP protocol. This is called *no coupling*, and both networks are peers. Mobile IP handles the mobility management. A *home agent/foreign agent* (HA/FA) entity is involved with this architecture, whereby the IP layer becomes aware of the agent advertisements of the mobility agent (HA/FA), and would do a binding update in certain lifetime [17].

2.7 Interconnection Between 3G-SGSN and WLAN Access Point by Emulating RNC

With this interconnection, the IEEE 802.11 WLAN network is connected to the UMTS CN via the Iu-PS interface. Figure 2.13 shows this heterogeneous network architecture.

IEEE 802.11 WLAN–based RAN is connected via an *interworking unit* (IWU), as shown in Figure 2.13, which is an RNC emulator. This is needed to exchange the packets between IEEE 802.11 WLAN network and UMTS. The function of the IWU is similar to an RNC in the (UMTS terrestrial RAN) UTRAN. It has to relay the Iu bearer service on the CN side to the distribution network [IEEE 802.3 *local area network* (LAN)] bearer service on the other side. The adapted UMTS bearer concept includes an appropriate location and mobility management for the terminals in the IEEE 802.11 WLAN coverage area. Due to the very small cell size of IEEE 802.11 WLAN systems, the access points are not directly connected to the UMTS CN. This reduces the signaling load caused by mobility and location management. A distribution network connects the WLAN access points and enlarges the coverage area of this radio access form. The IEEE 802.11 WLAN is treated as a routing area associated with the 3G-SGSN. Thus, the WLAN looks like an RNC to the UMTS network. A user, whether connected to the UMTS network or the WLAN, will always be treated as a UMTS user. The UMTS

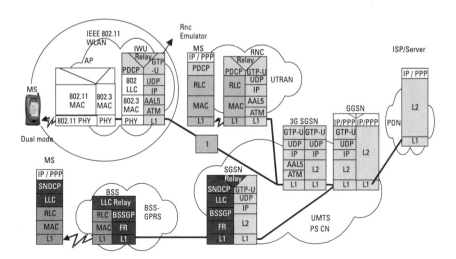

Figure 2.13 Interconnection between IEEE 802.11 WLAN AP and 3G-SGSN through an IWU that emulates RNC.

mobility management will have to maintain information about the user even when it is connected to a WLAN network. The IWU entity is the RNC emulator, which is presented in Figure 2.14.

The RNC emulator could be a LAN entity or a UMTS entity implemented in the networks. The LAN entity avoids encapsulation for routing to the UMTS network. For this type interconnection, a dual IEEE 802.11 WLAN/UMTS mode mobile station is required to use both networks, as shown in Figure 2.15.

The intertechnology roaming arises when the user is connected to the WLAN network. For this interconnection, the users have to interface to the UMTS *Packet Data Convergence Protocol* (PDCP) network through the RNC emulator. UMTS-specific protocol such as is on top of the IEEE 802.11 MAC and the PHY layers implemented. UMTS-related signaling protocols are carried out between the protocols in the *mobile station* (MS) and the RNC emulator. The RNC emulator is a black box that hides WLAN-specific features from the UMTS network. The IP protocol is used to transfer packet-switched data over the Iu interface as well as in the CN. The *GPRS Tunneling Protocol for UMTS* (GTP-U) on the top of this transport IP layer provides a tunneling service through the CN until the access network encapsulates the user data. Hence, if IP packets are transmitted on user level, two IP layer exist in the packet-switched architecture.

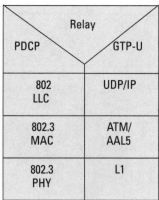

IWU as an RNC
emulator being a
UMTS entity

IWU as an RNC
emulator being a
WLAN entity

Figure 2.14 Protocol stack of the RNC emulator is a UMTS and WLAN entity.

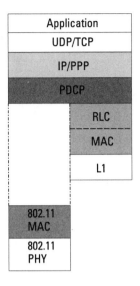

Figure 2.15 IEEE 802.11 WLAN/UMTS dual mode protocol stacks architecture of mobile station.

The IEEE 802.11 WLAN coverage area is represented as one routing area for the CN. If the mobile node leaves or enters a routing area, an update message is sent to the CN of UMTS. Hence, the 3G-SGSN can simply distinguish the different RANs via the routing areas. Running IP sessions are not interrupted because the IP address of a terminal is not changed. The procedure is completely transparent to the user. However, if a mobile leaves the IEEE 802.11 WLAN coverage area, the service quality will degrade, especially for those sessions that made use of the high throughput capabilities of IEEE 802.11 WLAN system.

The current UMTS approach foresees that within the CN, *differentiated services* (DiffServ) are used on the transport IP level to differentiate between different traffic classes. This approach can be mapped quite easily on both the IEEE 802.11 WLAN distribution network and the IEEE 802.11 WLAN bearer. If switched Ethernet implements the IEEE 802.11 WLAN distribution network, the DiffServ classes can be mapped onto IEEE 802.1p priorities and then to IEEE 802.11 WLAN MAC connections, and vice versa. Figure 2.16 shows the UMTS bearer concept [18] with IEEE 802.11 WLAN access integrated. The UMTS bearer is not changed with respect to the different radio interfaces. The RAB must be adapted to the new, underlying *distribution network* (DN) bearer and the IEEE 802.11 WLAN bearer.

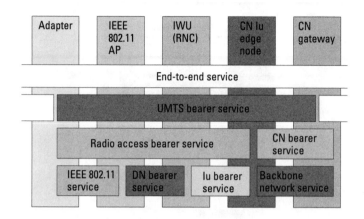

Figure 2.16 Adapted UMTS bearer concept using IEEE 802.11 WLAN bearer.

2.7.1   Pros and Cons of Emulating RNC

The main advantage of this interconnection together with the dual mode WLAN/UMTS protocol stack on MS, as shown in Figure 2.16, is that the mobility management, roaming, billing, and location-related issues are taken care of by UMTS network. *Subscriber identity module* (SIM) and *UMTS SIM* (USIM)–based authentication of a subscriber for WLAN offers a 2G/2.5G and 3G operator the following benefits:

- The WLAN subscriber credentials are of identical format to 2G/2.5G or 3G and therefore easier to integrate subscriber into the current HLR. Therefore, all existing roaming capabilities and settlements are inherited from GSM.

- The security level offered by WLAN will be identical to that of GSM/GPRS/UMTS. GSM SIM-based security is based on a challenge-response mechanism. It offers better tamper resistance because SIM runs an operator-specific confidential algorithm, which takes a 128-bit *random number* (RAND) and a secret key, Ki, stored on the SIM as an input to produce a 32-bit response (SRES) and a 64-bit data encryption keys, Kc(n), as an output. Kc(n) are never sent over the air, nor are they used in calculation for message authentication code for RAND and SRES. Kc(n), together with *international mobile subscriber identity* (IMSI) and Ki, are used by the network and the client to calculate independently the key K that will be used for the encryption of data over the air interface. So the only data exposed over the air interface are the random numbers, K(n).

Strong security provided in the UMTS network and QoS for real-time services may now be provided over WLAN, thereby resolving the drawbacks of current IEEE 802.11a/b/g WLAN threats. Minimum changes are required in the UMTS network, and this will create a master-slave relationship between UMTS and WLAN as discussed in Section 2.6. This is not optimal. Using UMTS PDCP frames over WLAN may create bottlenecks. In this scenario, the UMTS backbone may be a bottleneck for the WLAN traffic. WLAN data rates with 11 Mbps with IEEE 802.11b and 54 Mbps with IEEE 802.11a would be degraded to the speed of the UMTS terminal (2 Mbps). This type of interconnection requires modifications to standard WLAN terminals, which in turn would make them more expensive. Two attractive WLAN components (i.e., speed and price) would be lost in this type of connection.

2.8 Interconnection Between GGSN and WLAN Access Point by Emulating 3G-SGSN

As an alternative to the interconnection of WLAN to 3G-SGSN, it can also directly be connected to the GGSN of the UMTS network. Figure 2.17 shows the interconnection of WLAN to the GGSN of the UMTS network.

Architecture in Figure 2.17 is a modification to that of Figure 2.13 in that the interface between WLAN and UMTS is now via a 3G-SGSN–like device, which is called a 3G-SGSN emulator. The protocol stack is very similar to the one depicted in Figure 2.14. In this interconnection the Iu interface and the protocols between 3G-SGSN and 3G-GGSN are not used, and the functions supported by those protocols are thus not available. It is possible to bypass some of the RNC-related functions by using a 3G-SGSN emulator, and mobility management is again handled by UMTS.

2.8.1 Pros and Cons of Emulating 3G-SGSN

The disadvantage of the RNC emulator–based interconnection is the master-slave situation; bottlenecks and inefficient routing also exist in this interconnection type architecture. However, the advantage of this interconnection architecture is that some overhead caused by available but not needed functionality is avoided. In this architecture, the IWU requires the

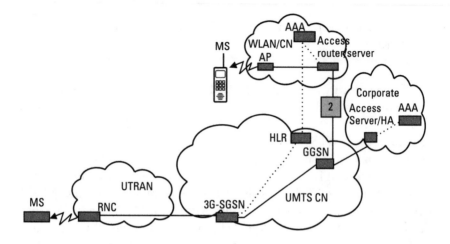

Figure 2.17 Interconnection between IEEE 802.11 WLAN AP and GGSN through an IWU
by emulating 3G-SGSN.

adaptation between WLAN and UMTS packet formats. If the 3G-SGSN
emulator does this adaptation, the GGSN could remain unaltered. In this
interconnection architecture, the GGSN throughput might become a problem if the GGSN capacity is not designed to fulfill the growth of traffic. If
the adaptation is done by the GGSN, taking into account the increased need
for bandwidth, the speed of WLAN could be exploited in full.

2.9   Interconnection Between UMTS and WLAN Through Virtual Access Point (VAP)

The VAP reverses the roles played by the UMTS and WLAN in the first two
interconnection architectures. Here, the WLAN is the master network, and
UMTS is the slave network. Figure 2.18 depicts the architecture of this interconnection type.

The difference of this interconnection compared to the other architectures is the existence of a VAP instead of RNC/3G-SGSN emulators. This is
the duality of the RNC/3G-SGSN emulators. Mobility is managed according to the IEEE 802.11 WLAN and *Interaccess Point Protocol* (IAPP) specifications by the WLAN standard. The intertechnology roaming that the
WLAN observes is between different access points in the extended service set
and the VAP that appears as yet another access point to the IEEE 802.11

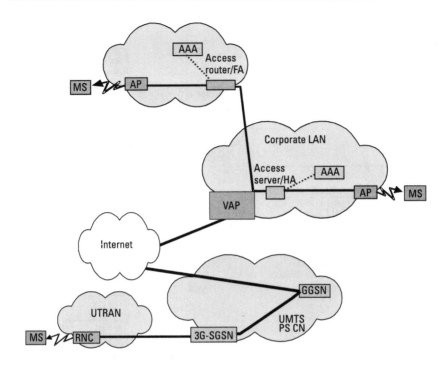

Figure 2.18 VAP-based interconnection to the UMTS.

WLAN. From the WLAN point of view, the entire UMTS network appears as a basic service set or a picocell associated with another access point (in this case, VAP). The function of the VAP is to communicate with MSs connected through UMTS, de-encapsulate their packets, and transmit them on the LAN. After this is done, the packets will reach the final destination through the router attached to the LAN. The protocol stack architecture is a modified version of that in Figure 2.13, only the VAP entity protocol stack is placed after the GGSN protocol stack. The protocol stack of the VAP entity is presented in Figure 2.19(a). The VAP-based interconnection also requires some modification on the MS protocol stacks, which is presented in Figure 2.19(b).

In this interconnection, the VAP becomes an adopted unit in the user plane protocol of the UMTS architecture. For this reason, the IEEE 802.11 MAC protocol is implemented on top of the protocols of the UMTS GGSN part. This is done for both the MS and the VAP entity. From the GGSN part

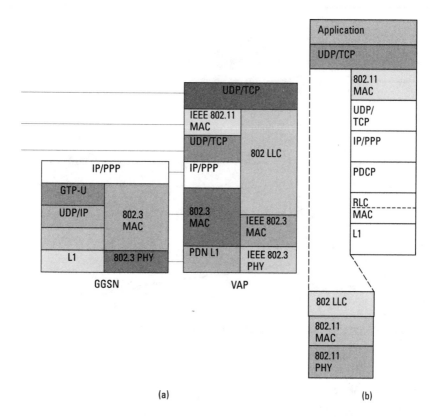

Figure 2.19 (a) Protocol stacks architecture associated with the VAP-based interconnection; and (b) protocol stacks architecture for the MS associated with the VAP-based interconnection.

of the UMTS network, all protocols up to GTP-U level will be mapped onto IEEE 802.3 MAC so that the WLAN network sees the VAP as an access point. On the VAP side, the UDP/TCP protocol is on the top of the stack. The 802.11 MAC protocol that is implemented in the MS is a level below the UDP/TCP. Hence, the 802.11 MAC protocols and the protocols below it are mapped to the 802 *logical link control* (LLC). The 802.3 MAC protocol in the GGSN part is mapped onto VAP below the IP/*Point-to-Point Protocol* (PPP) protocol in the VAP protocol stack.

2.9.1   Pros and Cons of VAP-Based Interconnection

This type of interconnection does not have significant advantages over the other four types of interconnections. It is not clear how the VAP will operate with the regular 802.11 WLAN access points. From Figure 2.19, it is clear that over the UMTS air interface, each packet will have twice UDP/TCP and twice IP/PPP headers. In addition, there will be an IEEE 802.11 MAC header along with the UMTS-related headers. The overhead of packets on the low bandwidth air interface is quite large, which makes this interconnection inefficient.

2.10 Interconnection Between UMTS and WLAN Through Mobility Gateway

The interconnection architecture is presented in Figure 2.20. An intermediate server (mobile proxy) is placed on either the UMTS or the IEEE 802.11 WLAN sides and the *mobility gateway* (MG) will handle the routing and mobility issues.

When an MS is attached to an access point, the communication path between the MS and a *correspondent host* (CH) on the Internet will be 1a-3. The CH-MS communication path will be 4-2a. When the MS is on the UMTS network, this path will be 1b-3, and the reverse path is then 4-2b. It

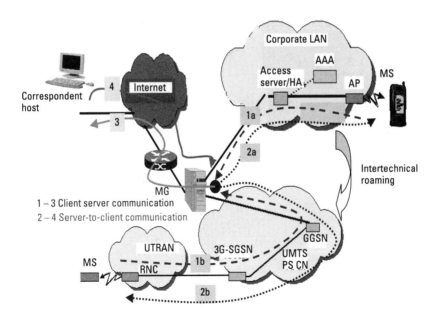

Figure 2.20 MG-based interconnection between IEEE 802.11 WLAN and UMTS.

should be observed that the segments 3 and 4 in both paths do not change regardless of where the MS is located. Only the links 1 and 2 will be continually changing, depending on the movement of the MS. Clearly, the communication between MS and the proxy server alone is subject to change while maintaining the proxy-CH connection unchanged supports mobility. On protocol stacks architecture of the WLAN network nothing changes. The protocol architecture of the UMTS network is a modification of which the MG is placed next the GGSN part of the protocol stacks of the UMTS. On the MS, there will be some protocols adopting if the user want roaming between both networks when he or she is still in one of the network. So this function does require a dual-mode stack implementation on the MS. Figure 2.21 shows the protocol stack architecture for both the MG entity and the MS.

2.10.1 Pros and Cons of Interconnection Between UMTS and WLAN Through MG

There are several advantages to employing proxy architecture for inter-technology roaming. The proxy architecture is scaleable. There is the possibility of further minimizing the encapsulation and routing inefficiencies with

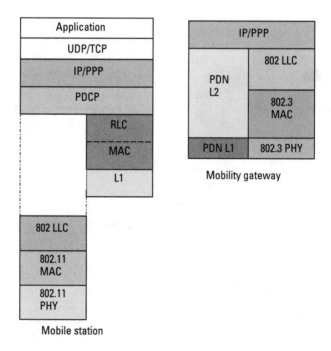

Figure 2.21 Protocol stacks corresponding with the MS and the MG.

mobile IP. However, the real reduction in overhead may not be very significant due to the need for additional control protocols. If the proxy is under the control of the same organization that owns the mobile hosts/station, it is possible to configure the proxy to support the peculiar needs of its population of mobile hosts. An optimized protocol may be run between the mobile and proxy depending on the link being employed. The proxy can manage resource-poor connections more efficiently. For example, it can drop structured data such as e-mail headers and frames in a *Motion Picture Expert Group* (MPEG) stream selectively or drop unstructured data.

Proxies are already in place in many organizations as firewalls or Web caching servers. These may be reused for mobility management and intertechnology roaming. Proxies can be used for logging the characteristics of connection, the details of which may be usefully employed in various applications including accounting and management.

The main disadvantages of the proxy architecture are as follows. First, the architecture is not standardized, and therefore it requires proprietary protocols for intertech roaming. Second, the performance of a proxy is poor because a significant latency is added to the client-server communication path. Potentially, the end-to-end semantics of the transport protocol may also be violated. If a single proxy is employed and it fails, it may result in the failure of the entire network—there is a need to have some fault tolerance. Third, there is the issue of developing protocols for mobility management with the proxy architecture. There are some more issues that are open. The placement and number of proxies to be employed may depend on the situation. It is preferable to have the proxy connected to the last links of each services that the MS may use so that it can gather information about the quality of each of the last links. However, the ownership of such a proxy will be contentious. The number of proxies that have to be placed for optimum performance is also subject to network conditions.

2.11 Interconnection Between UMTS and WLAN Based on Mobile IP

The interconnection architecture related to the mobile IP is presented in Figure 2.22. Mobile IP is employed to restructure connections when an MS roams from one network to another. Outside of its home network, the MS is identified by a *care of address* (COA) associated with its point of attachment and a colocated FA that manages deencapsulation and delivery of packets.

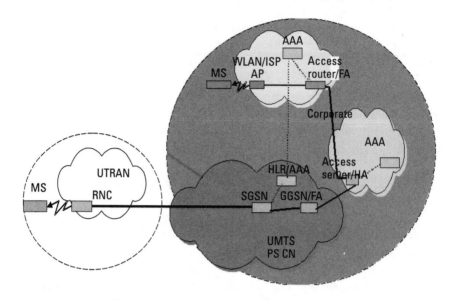

Figure 2.22 Interconnection architecture between WLAN and UMTS based on mobile IP.

The MS registers its COA with an HA. The HA resides in the home network of the MS and is responsible for intercepting datagrams addressed to the MS's home address as well as encapsulating them to the associated COA. The datagrams to an MS are always routed through the HA. Datagrams from the MS are relayed along an optimal path by the Internet routing system, though it is possible to employ reverse tunneling through the HA.

The required dual-mode MS protocol stack is the same as in Figure 2.21. It is clear that both networks are peer networks and that the functionality of the HA/FA exists at the IP layer.

2.11.1 Pros and Cons of Interconnection Between UMTS and WLAN Based on Mobile IP

The advantage of this interconnection is that it is based on the mobile IP, which makes the IP address mobile. The same IP address is used, which solves the multiple address problems. To solve the packet duplications due to the lifetime of the routers, some conventions on both IEEE 802.11 WLAN and UMTS networks are needed. The databases of both networks may need to communicate to overcome packet duplication.

The main disadvantage of Mobile IPv4 is the triangle routing. This could be overcome with the mobile IP with the use of optimized routing. This is important for real-time applications like video or audio transmission.

2.12 Handover Between IEEE 802.11 and UMTS

The motivation intertechnology (vertical handover) for the hybrid mobile data networks arises from the fact that no one technology or service can provide ubiquitous coverage, and it will be necessary for a mobile terminal to employ various points of attachment to maintain connectivity to the network at all times. There is a clear difference between the two types of handover—namely, horizontal and vertical handover. Horizontal handover refers to handover between node Bs or access points that are using the same kind of network interface. Vertical handover refers to handover between a node B and an access point (or vice versa) that are employing different wireless technologies. In the case of a vertical handover, two distinctions are made:

- Upward vertical handover, which occurs from IEEE 802.11 WLAN access points with small coverage to an UMTS node B with wider coverage;
- A downward vertical handover, which occurs in the reverse direction.

A downward vertical handover has to take place when coverage of a service with a smaller coverage, as in WLAN service, becomes available when the user still has connection to the service with the UMTS coverage. An upward vertical handover takes place when an MS moves out of the IEEE 802.11 WLAN coverage to UMTS service when it becomes available but the user still has a connection to the IEEE 802.11 WLAN coverage. In the case of the upward vertical handovers, the mobile station/host decides that the current network is not reachable and hands over to the higher overlay UMTS network, when several beacons from the serving WLAN service are not available. It instructs the WLAN to stop forwarding packets and routes this request via a mobile IP registration procedure through the UMTS CN. When it is connected to the UMTS network, the MS listens to the lower layer WLAN access point. If several beacons are received successfully, it will switch to the IEEE 802.11 WLAN network via the mobile IP registration process. The vertical handover decisions are thus made on the basis of the presence or absence of beacon packets.

2.13 Handover Aspects Between IEEE 802.11 WLAN and UMTS Based on Mobile IP

Handover is the mechanism by which ongoing connection between MS and CH is transferred from one point of access to another point while maintaining the connectivity. When an MS moves away from an access point or from a node B, the signal level degrades and there is a need to switch communications to another point of attachment that gives access to the existing IEEE 802.11 WLAN network or UMTS network. Handover mechanism in an overlay UMTS and underlay WLAN network could be performed so that the users attached to the UMTS just occasionally check for the availability of the underlay WLAN network. A good handover algorithm is needed to make the decision when to make handover in order to avoid an unnecessary handover (i.e., the ping-pong effect). The handover algorithm is beyond the scope of this chapter. This section discusses the handover procedure and the mechanism from WLAN to UMTS, and vice versa, based on the *received signal strength* (RSS) metrics. It means that the handover initiation or the handover triggering is sensitive to these signals. Figure 2.23 shows the handover procedure by transition from one network to another.

An MS moving from the WLAN network coverage may suddenly experience severe degradation of service and will have to perform handover very fast to maintain the higher layer connection. The following stages occur when an MS moves away from the coverage of WLAN within the UMTS coverage:

1. The signals received from the access point in WLAN network is initially strong and the MS is connected to the WLAN network, which is also the home network of the MS and the HA in this network.

2. The signals from the access point become weaker when the mobile moves away. The mobile scans the air for another access point. If no access point is available, or if the signal strengths from the available access point are not strong enough, the handover algorithm uses this information along with other possible information to make a decision on handing over to the higher overlay UMTS

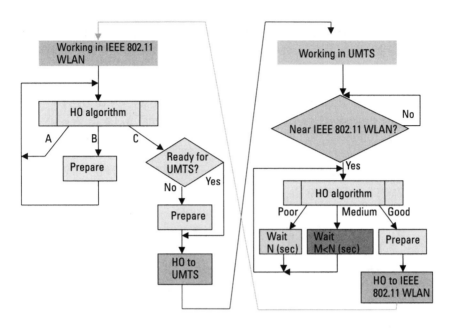

Figure 2.23 Handover procedure between WLAN and UMTS.

network. Connection procedures are initiated to active the UMTS PCMCIA card.

3. The handover algorithm in the MS decides to dissociate from the WLAN and associate with the UMTS network.

4. The FA is activated and used by the MS dual PCMCIA card and the mobile IP protocol, and the MS gets a COA due to visiting the UMTS network as a foreign network.

5. The HA in the WLAN is informed about the new IP address through a mobile IP registration procedure, and it does proxy *Address Resolution Protocol* (ARP) and intercepts the datagram. The HA encapsulates the datagram and tunnels any packets arriving for the MS to the FA of the UMTS networks. At the end of the delivery, the MS will deencapsulated and get the datagram.

In this case, the handover algorithm determines that there is no local coverage available via WLAN and handover must be performed to the UMTS network, assuming that a UMTS service is always available to the MS.

Once the MS is attached to the UMTS, it constantly monitors the air at repeated intervals to see whether or not a high data rate WLAN service is available. As soon as such a service becomes available, the handover algorithm should initiate an association procedure to the newly discovered access point. Figure 2.24 illustrates the handover procedure from WLAN to UMTS.

The procedure for this reverse handover from UMTS to WLAN network is as follows:

1. The signal from the WLAN access point is initially not detected.

2. The MS then detects a beacon, which indicates that the underlay WLAN network has become available.

3. The handover algorithm decides on making a handover from UMTS to the WLAN network.

4. The FA in the UMTS network is deactivated, the mobile IP is updated, and the home IP address is used.

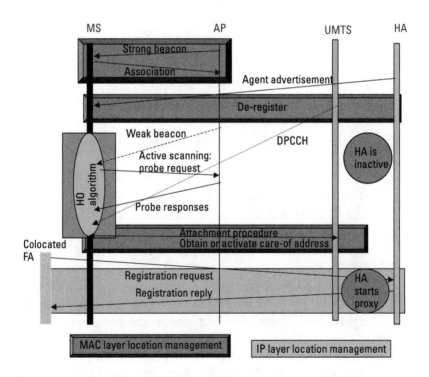

Figure 2.24 Messages and signaling of the handover procedure from IEEE 802.11 WLAN to UMTS.

5. The HA in the WLAN network is instructed by the MS to no longer do a proxy ARP on its behalf through the mobile IP protocol.

2.14 Conclusions and Future Directions

After introducing the 3G networks and standards, this chapter presented five interconnections and seamless intertechnology handover dealing with interworking aspects between IEEE 802.11 WLAN and UMTS. The concepts discussed focused on the network layer and the link layer in order to minimizing effects on existing networks and technologies, especially at the lower layers such as MAC and PHY.

Based on the arguments presented for the given interworking approaches, mobile IP-based interconnection architecture is selected as the most suitable solution. IP layer mobility management provides an efficient way to interconnect heterogeneous packet-oriented networks. Interworking cannot be handled within a proprietary protocol of one network technology; it has to be handled either in the existing layers above or a new layer has to be added solely for the purpose of handling intertechnology roaming. In either case, there is a need for modification of existing protocols, at least between the MS and a network entity that handles the mobility.

Finally, intersystem handover is a topic that will become more and more important in the evolutionary path towards UMTS and 4G wireless infrastructures. Mobile IP-based intertechnology roaming is an evident step on this path. The fast handover required for real-time services have to be studied further within a platform.

Current telecommunication and computer networks are on the verge of providing mobile multimedia connectivity, where nomadic users would have ubiquitous access to remote information storages and computing services. As an evolutionary step towards the 4G mobile communications, mobility in heterogeneous IP networks with both UMTS and IEEE 802.11 WLAN systems is seen as one of the central issues in making the 4G of telecommunication networks and systems.

References

[1] Mehrotra, A. K., *GSM System Engineering*, Norwood, MA: Artech House, 1997.

[2] Prasad, N. R., "GSM Evolution Towards Third Generation UMTS/IMT2000," *ICPWC'99*, Jaipur, India, February 17–19, 1999.

[3] Prasad, N., and A. Prasad, *WLAN Systems and Wireless IP for Next Generation Communication*, Norwood, MA: Artech House, 2002.

[4] IEEE, "IEEE standard for WLAN MAC & PHY specifications," 1997, p. 445.

[5] Ojanperä, T., and R. Prasad, *WCDMA: Towards IP Mobility and Mobile Internet*, Norwood, MA: Artech House, 2001.

[6] Ojanperä, T., and R. Prasad, *Wideband CDMA for Third Generation Mobile Communications*, Norwood, MA: Artech House, 1998.

[7] Prasad, R., W. Konhäuser, and W. Mohr, *Third Generation Mobile Radio Systems*, Norwood, MA: Artech House, 2000.

[8] Prasad, N. R., "An Overview of General Packet Radio Services (GPRS)," *First International Symposium on Wireless Personal Multimedia Communications (WPMC'98)*, Yokosuka, Japan, November 4–6, 1998.

[9] 3GPP2, http://www.3gpp2.org.

[10] http://www.cdg.org.

[11] DaSilva, J. S., et al., "European Third-Generation Mobile Systems," *IEEE Comm. Mag.*, October 1996, Vol. 35, No. 10, pp. 68–83.

[12] Huber, J. F., D. Weiler, and H. Brand, "UMTS, the Mobile Multimedia Vision for IMT-2000: A Focus on Standardization," *IEEE Comm. Mag.*, 2000.

[13] 3GPP, http://www.3gpp.org.

[14] ETSI, "Requirements for the UMTS Terrestrial Radio Access System," UMTS 04-01 UTRA, June 1997.

[15] Chaudhury, P., W. Mohr, and S. Onoe, "The 3GPP Proposal for IMT-2000," *IEEE Comm. Mag.*, December 1999.

[16] ITU: http://www.itu.int/imt/2-radio-dev/rtt/index.html, July 1998.

[17] IP Mobility Support for IPv4: http://www.ietf.org/rfc/rfc3220.txt.

[18] http://www.3gpp.org, UMTS 23.07 V0.4.0, "Quality of Services Concept."

3

TCP/IP Protocol Stack

3.1 Introduction

Mobile communication networks are to an extent evolving in the same way as wired networks have, but over a much shorter time period. That is, first-generation analog networks were devised to offer users a national-level mobile telephone service, but without the possibility of interconnection with a network besides the telephone network. Second generation digital networks added the capacity of providing international-level services (i.e., becoming more open and, at the same time, offering data services, although with a low capacity). Third generation networks were deployed with the aim of dealing with voice and data services in equal measures, with a special emphasis on the interconnection of user terminals with the Internet. Finally, the networks beyond 3G may be heterogeneous networks in which, by means of vertical handover procedures, the internetworking of different infrastructures of wireless networks will be achieved. In this last step, the technology bringing together this heterogeneity will be the *Transmission Control Protocol/Internet Protocol* (TCP/IP), which is at the heart of the Internet [1–3].

Given that TCP/IP is without doubt the protocol pile that will support the different applications that will be executed in the future from fixed, mobile, and portable terminals, it is necessary to describe both IP and TCP to be able to evaluate their performance. The performance of the corresponding applications, when working over wireless platforms, is described more fully in Chapter 5.

The origin of TCP/IP can be dated to the end of the 1960s when the *Defense Advanced Research Projects Agency* (DARPA) of the United States detected the proliferation of computers in the military environment. Because the computers for military use, like civilian ones, were made by different manufacturers, the solutions provided for their communications were closed. That is, the communication between the products of different manufacturers was not possible. With the aim of remedying this situation, the U.S. Department of Defense asked the DARPA to define a common communications protocol that would facilitate interoperability among heterogeneous equipment. In this way, what we now know as the Internet was born, although then it was denominated *Advanced Research Project Agency Network* (ARPAnet), initially made up of only four packet switching nodes.

This chapter is organized as follows: Section 3.1 describes the IP protocol including fragmentation and addressing issues. The *Internet Control Message Protocol* (ICMP) and the *Address Resolution Protocol* (ARP) are introduced in Sections 3.3 and 3.4, respectively, while Section 3.5 explains the routing protocols. Mobility and the IP issues are discussed in Section 3.6 by presenting macro- and micro-mobility protocols. Last, but not least, the transport protocols are discussed in Section 3.7.

3.2   IP

The IP is the internetworking protocol that offers a service with the following characteristics [4, 5]:

- It is connectionless, so units of network layer data protocol, denominated *datagrams* in the IP context, are dealt with in an individual way from the source host up to the destination host.
- It is not reliable. The datagrams can be lost, duplicated, or disordered, and the network does not detect or report this problem.

As for the IP functionality, two basic functions are carried out by the IP: addressing and fragmentation. The IP entities use the IP address to transmit the datagrams toward their destination. The selection of the path to be followed is called *routing*. In the same way, the IP entities use particular fields of the header of the datagrams for fragmentation and reassembly when necessary.

The IP datagram has the structure shown in Figure 3.1.

Datagram header field	Datagram data field

Figure 3.1 IP datagram format.

The IP datagram data field transports the information belonging to the protocols of higher layers, while the IP datagram header field transports the information that the IP entities really handle and therefore determine the previously mentioned functionalities. The format of the IP header, shown in Figure 3.2, It contains a fixed part of 20 bytes and an optional part of between 0 and 40 bytes.

The version field allows the unambiguous coexistence, in the same network, of packets of different versions of IP; the version currently being used (corresponding to the structure of the datagram that we are seeing) is version 4.

The *HLen* field specifies the length of the header, given that it can vary due to the presence of optional fields. It is specified in words of 32 bits. The minimum length is five and the maximum is 15, the equivalent of which is 40 bytes of optional information. The length of the header must always be a

0			15 16		31
Version	HLen	Type of service		Total length	
Identification			DF MF	Fragment offset	
Time to live		Protocol	Header checksum		
Source IP address					
Destination IP address					
IP Options					
Data					

Figure 3.2 IP header format.

whole number of words of 32 bits. If the length of the optional fields is not an exact multiple of 32 bits, a padding is added to the end of the header.

The *type of service* field is used to indicate to the network the desired QoS [6]. This field, made up of 8 bits, has evolved over time in terms of its structure, although not in its fundamental aim, in such a way that it is now used to support what is known as differentiated services or DiffServ. To be exact, these 8 bits are assigned to two fields, one of 6 bits called *differentiated services codepoint* (DSCP), which allows the handling of up to 64 levels, and another of 2 bits of reserved use.

The *total length* field specifies the length of the whole datagram (including header) in bytes. The maximum value is 65,535 bytes.

The next four fields, *identification, don't fragment* (DF), *more fragments* (MF), and *fragment offset,* are related to the fragmentation of datagrams, which will be explained later.

The *time to live* (TTL) field allows the discarding of a datagram when it has made an excessive number of hops or has passed an excessive time traveling in the network and is presumed useless. It is a regressive counter that indicates the remaining lifetime of the datagram measured in seconds, in such a way that the datagram must be discarded if its value reaches zero. Its purpose is to avoid the production of loops due to a routing problem and to prevent a datagram from remaining indefinitely in the network. It is not trivial to precisely calculate the time that the datagram takes to travel between two gateways; therefore, in practice what is done is to subtract one from the TTL for each hop, and if the datagram waits more than one second at a gateway to subtract one for each second of waiting time. As the datagrams hardly ever stay at a gateway more than one second, in practice, this parameter works as a hop counter. If fragmentation takes place, the receiver host can retain datagrams for several seconds while it waits for all of the fragments to arrive. In this case, the host subtracts one from the TTL for each second waiting and can even discard datagrams for this reason. Values of TTL of 32 or 64 are habitual.

The *protocol* field specifies which transport-level protocol the datagram corresponds to. Table 3.1 shows some of the possible values of this field.

The *checksum* field is for detecting errors produced in the header of the datagram. It is obtained from the one's complement of the header taken in fields of 16 bits (including the optional fields if there are any). The sum of one's complement of all of these blocks gives the result. The checksum enables the protection of the datagram from an alteration in some field of the header, which could be produced, for example, by a hardware problem at a gateway. Note that the checksum only covers the header of the datagram, not

Table 3.1
Some Examples of the Protocol Field Values

Protocol field	Protocol	Description
0		(Reserved)
1	ICMP	Internet Control Message Protocol
3	GGP	Gateway-to-Gateway Protocol
4	IP	IP in IP (encapsulation)
5	ST	Stream
6	TCP	Transmission Control Protocol
17	UDP	User Datagram Protocol
46	RSVP	Reservation Protocol
89	OSPF	Open shortest path first
255		(Reserved)

the data. The checksum field must be calculated in each hop, because the TTL, at least, changes. At gateways with a lot of traffic, the recalculation of the checksum is a drawback from the point of view of performance.

The *source address* and *destination address* fields correspond to IP addresses in accordance with the format we will see later.

The optional fields of the header are not always supported at the gateways and are rarely used; of these, the following can be highlighted:

- *Record route.* This option requests each gateway that the datagram goes through to write its address in the header in order to provide a trace of the route followed for tests or diagnosis of problems.

- *Timestamp.* This option acts in a similar way to a record route, but as well as noting the IP address of each gateway. It notes in another 32-bit field the instant at which the datagram passes through the gateway.

- *Source routing.* It allows the emitter to specify the route that the datagram must follow to the destination. There are two variants. Strict source routing specifies the exact route hop by hop in such a way that if in any case the route specified is not possible for any reason, an error will be produced. Its functionality is equivalent to that of bridges with routing from the source. With loose source routing, the gateways the datagram should pass through are specified, but the network has freedom to use other unspecified intermediate gateways when it is considered necessary.

The use of optional IP header fields generally leads to problems of performance, given that the implementations of the gateways optimize the code for normal situations (i.e., for datagrams without optional fields).

3.2.1 Fragmentation

The IP-level datagram must be encapsulated in a lower network level packet to travel in the network. The ideal case could be considered to be that in which a network header is added to the datagram and is sent through the same network. In the real case, the size of the packet of the network can vary greatly and can be smaller than the size of the IP datagram, so it is necessary to fragment the IP datagram in several pieces to send it through the network. This is where it is necessary to introduce the concept of *maximum transfer unit* (MTU), which is the maximum size of a packet that can be transferred by a specific network.

To approach this problem, the size of the IP datagram could have been limited to the smallest MTU of the networks, but this system would be highly inefficient in networks where a larger sized packet were admitted.

Really, the solution consists in, on the one hand, choosing a suitable packet size, and, on the other, designing how to divide the datagram in fragments when it is necessary.

The gateways undertake this fragmentation, and the rules for the fragmentation are as follows:

- The size of the resulting fragments must be a multiple of an octet so that the data displacement records, offset, within the datagram are done correctly.

- The sizes of the fragments are freely chosen (i.e., the initial size does not have an influence and the resulting fragments can be of differing lengths).

- The gateways must accept datagrams with a greater size than that of the network they are connected to. This is so larger datagrams can be admitted to the network, although they must be fragmented to travel in it.

- The hosts and gateways must handle datagrams larger than 576 octets.

Each fragment has the same format as the original datagram, which means that the fragments generated will have a header that is the same as the original datagram, except for a series of differences explained next. The data of each fragment will be made up of fragments of the original datagram's data. The size of each fragment will be limited by the MTU of the network where they are going to be sent.

We will now look at the differences between the header of the original datagram and those of its fragments. If we consider that the original datagram has been divided into N fragments, the first $N-1$ fragments will have the MF flag set to 1 and the fragment N will have this flag set to 0, indicating that there are no more fragments. The fragment offset field is used to indicate, when the datagram is a fragment, which position in the original datagram the data are located. The fragmentation is always done in multiples of eight bytes, which is the elemental unit of fragmentation. Thanks to this, the field requires only 13 bits instead of the 16 that would be necessary if bytes were counted. Of the 3 bits saved, two are used in the DF and MF flags and the third is not used. Note that the entity that must reassemble the datagram would have no way of knowing its total length if it were not for the last fragment with its fragment offset and total length fields, which permit it to know the total length of the datagram. In both cases, the identification field, along with the datagram source address, allow the gateway/destination to know which arriving fragments belong to which datagrams.

At this stage we ask the following question: Who must reassemble the fragmented datagram? There are two options: The first is that the first gateway receives the fragments at the next routing point of the Internet; the second option is that the reassembly is done by the destination host of the original datagram.

The disadvantages of the second option, which is implemented in practice, are first that the fragments can cross networks with a greater MTU without taking advantage of its capacity to transmit larger sized packets. Second, it is obvious that if one fragment of a datagram is lost, the whole datagram is lost.

In spite of this, the system works well and has the advantage that the intermediate gateways do not need to store and reassemble packets, which means the gateways are simpler and require less capacity.

Finally, there is another interesting aspect to deal with related to the way the MF bit is propagated if it is necessary to further fragment the datagram's fragments. It is done in a very simple way. For all of the fragments with the MF bit at 1, all its fragments have this bit set at 1. On the contrary, when a fragment is received with the MF bit at 0, this bit is set at 1 for all the fragments except the last one, which is set to 0.

Once the fragments arrive at a destination address, they must be reassembled. For this, the IP software must have a system implemented such that:

- It enables the rapid insertion of the fragment in the group. As the fragment offset field of the fragments refers to the original datagram, space can easily be reserved for each fragment.

- It must have an efficient test of the whole datagram.

- It must quickly eliminate the stored fragments if they surpass the maximum time of reassembly.

If at a particular moment the fragment storage capacity is exceeded, all of the fragments are discarded. This is done because one lost fragment makes it impossible to obtain the whole datagram; in this way, space is liberated for other datagrams to be reassembled.

3.2.2 Addressing

Each network interface of each host or gateway must be identified with an IP address, and these addresses must not be repeated. The IP addresses are structured in a hierarchical way, as they are made up of numbers of 32 bits, and one part identifies the network, *netid,* and the other identifies the host within the network, *hostid.*

It was impossible to predict how the networks would be configured when the Internet was created, and it was necessary to adapt the format of the IP address to all types of current and future networks. For this purpose, five address formats were defined corresponding to a series of classes, from class A to class E.

Classes A, B, and C are currently used, whereas class E is reserved for future use. Figure 3.3 shows the format for each of these classes. From the figure, it can be deduced that class A has a great capacity to address hosts within the subnetwork and so is an appropriate class for large networks, while class C has a much smaller capacity for addressing hosts within this same subnetwork and so is suitable for subnetworks with up to 256 hosts. On the contrary, as only seven bits have been assigned to netid, in class A, the number of addresses of this class is low, exactly the opposite of what happens with class C addresses.

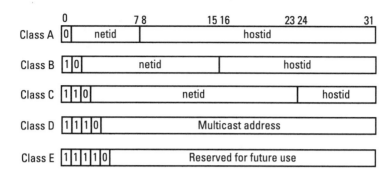

Figure 3.3 IP address classes.

Finally, class D belongs to what is called *multicast address* and is used to transmit datagrams simultaneously to a group of hosts making up a multicast group.

In practice, the network address makes use of a notation corresponding to the decimal value of each octet separated by points. Table 3.2 shows the ranges of valid values for the different classes expressed in decimal notation.

There is a series of addresses that by convention have a special meaning, which means that they are reserved. They have all the bits at 0 or all the bits at 1 of the netid and hostid fields. Thus, if the hostid field has all its bits at zero, it is understood that the address corresponds to the network and not to a specific host. On the contrary, if this field has all of the bits at 1, it indicates that the datagram should be broadcast to all of the hosts connected to the subnetwork. This is denominated *broadcast address*.

The network addresses are assigned by the *Network Information Center* (NIC). Generally, portions of the network are assigned to organizations, and these organizations are left to handle addresses internally. When working in a

Table 3.2
Decimal Value for the Different Address Classes

Address Class	Address Ranges
A	0.0.0.0–127.255.255.255
B	128.0.0.0–191.255.255.255
C	192.0.0.0–223.255.255.255
D	224.0.0.0–239.255.255.255
E	240.0.0.0–255.255.255.255

network isolated from the exterior, free addressing can be used, but it is advisable to have a network number set by the NIC in case of connection with the exterior.

Given that the human handling of the IP addresses is not simple, each address has an associated name. Thus, for example, the IP address 193.144.186.2 is assigned the name *moon.tlmat.unican.es*. This assignation of names is hierarchical; it is based on a scheme of the form *system.subdomain...subdomain.subdomain.domain*. There is a distributed application, *Domain Name System* (DNS), permitting the resolution of associations among names and IP addresses.

To finish this section, some of the problems associated with the addressing scheme are commented:

- If a host transfers from one network to another, it must change its IP address completely.
- When a class C surpasses 254 hosts, it must change to class B, which obliges the readdressing of all of the hosts of this network.

3.3 ICMP

As has been mentioned, IP provides a service for datagram transmission without guaranteeing the arrival at the destination. Furthermore, as it is not a connection-oriented protocol, there is no coordination among the origin, destination, and all of the intermediate gateways intervening in the communication. Therefore, it is helpful to have a mechanism for informing the source of the message about the cause of a loss or any other event associated with a datagram. The ICMP permits this utilization of the data part of a standard IP datagram [7]. It is important to note that including information in the data field of a datagram should be considered not a higher level protocol but an integral part of IP, and should be included in all IP implementations.

The errors detected by the ICMP are varied. One is *host unreachable*, which informs about the impossibility of reaching the destination because it cannot be found either because it is disconnected or because there is an error in the address. Others are *time exceeded*, which reports that the TTL field of the datagram has been decreased down to zero without reaching its destination and *source quench*, which indicates a state of congestion in the gateway. These messages can be used, for example, so that the local host stops the transmission when it detects congestion at a gateway in the network.

It is important to note that ICMP is a protocol permitting the gateways to report an error, but it does not provide mechanisms for its correction. In fact, ICMP does not specify what to do when an error is produced. It is even possible that the source of the message can do nothing about it (typically due to congestion at an intermediate gateway). Only the source of the message is informed, and not the intermediate gateways, is because usually an IP datagram includes only origin and destination addresses and not the complete route followed. This makes it impossible for a gateway to know the path followed by a specific datagram, especially if the routing mechanism used updates the tables dynamically.

3.3.1 ICMP Message Format

The ICMP message is encapsulated, as was explained earlier, within the data field of an IP datagram. To distinguish these messages from an ordinary IP datagram, the protocol field of the IP datagram contains the value 1, indicating that the data field in turn contains an ICMP message. This encapsulation is shown in Figure 3.4.

The ICMP header in turn contains three fields:

- *Type,* identifying the type of message;
- *Code,* providing information about the type of message;
- *Checksum,* containing the checksum of the whole ICMP message. The algorithm used is the same as for the IP protocol.

As for the data field, its content depends on the type of message. However, it usually contains two pieces of information: the header of the IP datagram that caused the error, and the first 64 bits of the data field of the same datagram. As the IP header is included, the source can determine which

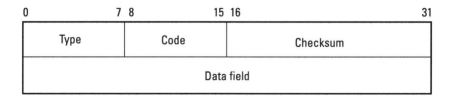

Figure 3.4 ICMP message format.

datagram provoked the error. In the first 64 bits of the datagram, the header of the messages of a higher level protocol, usually TCP, can be found. This can be used to determine which protocol and which application is responsible for the datagram in question.

3.4 ARP

The ARP is intended to solve the problem derived from the fact that the Internet is a virtual network working on physical networks [8]. When a datagram must travel from one machine to another (to the destination host or to a gateway) located in the same physical network, the physical layer that exists below the Internet encapsulates the datagram with the frame of the network itself, which must include the physical address of the machine. The problem consists in that this physical address can be unknown and must be obtained from the known IP address.

There are two basic methods to obtain the physical address.

- *Direct mapping.* Direct mapping, the most trivial way, allows the physical address to be found from the IP address. Direct mapping can be used when the physical address fulfils one of the following two conditions: either it can be chosen freely or it is sufficiently small. In the first case, it is enough to assign each machine a physical address that coincides with the part of the IP address that identifies the host. In this case, the determination of the physical address is trivial. In the second case, it is enough to have a simple function that maps each IP address to a physical address.

- *Dynamic mapping.* Dynamic mapping requires a specific protocol such as the ARP to find out the physical address. When the physical address is fixed, or when it is made up of a large number of bits, direct mapping is unviable. This is the case of Ethernet networks, in which the address of a machine is made up of 48 bits fixed by each manufacturer and cannot be modified. A simple solution is to have a table with the correspondences among addresses, but it must be modified every time a new machine is added to the network. The ARP protocol provides a mechanism to resolve this problem in a dynamic way with the only condition that the network has the capacity of broadcasting.

The ARP basically works by sending all of the machines in the physical network a datagram with a special format including the IP address. All of the machines will receive the message, but only the one whose IP address coincides will send a reply with its physical address. From this moment on, the physical address of the host is known.

This initial method can have some improvements. The first consists of maintaining a cache memory with the last physical addresses obtained. Thus, when a physical address is needed, the cache memory is consulted first, sending the ARP message only if there is no correspondence. Another improvement is based on the fact that if a machine A needs the physical address of B, it is likely that B will need the address of A in the near future. It is enough for A to include its physical address in its request message to B. Furthermore, as all of the machines in the network receive this request, they can take advantage by storing the physical address of A in their corresponding caches for future communications.

3.4.1 ARP Message Format

As the ARP is strongly dependent on the addressing of the physical network, there is no fixed format for the ARP messages. In general, the ARP message is encapsulated within the frame used by the network infrastructure, as shown in Figure 3.5.

To identify a frame as the carrier of an ARP message, the type field of the frame header takes a specific value. For example, for Ethernet, the type field takes the hexadecimal value 806.

The ARP message is designed for any type of physical network, so its length is variable depending on the type of addressing. Figure 3.6 shows an ARP message including the Ethernet hardware addresses.

The hardware type field indicates the type of physical network (for Ethernet it takes a value of 1). The protocol type field specifies the protocol of the network used, which in the IP case takes the hexadecimal value of 800.

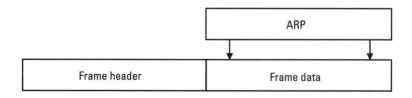

Figure 3.5 ARP message encapsulated in the corresponding frame.

0		15 16	31
Hardware type		Protocol type	
HLen address	IPLen	Operation	
Sender hardware address (bytes 0...3)			
Sender hardware address (bytes 4-5)		Sender IP address (bytes 0-1)	
Sender IP address (bytes 2-3)		Target hardware address (bytes 0-1)	
Target hardware address (bytes 2...5)			
Target IP address (bytes 0...3)			

Figure 3.6 ARP packet encapsulated in an Ethernet frame.

The operation field indicates whether the packet contains an ARP request or an ARP reply.

The HLen and IPLen fields specify the length of the hardware (physical) and IP addresses, respectively. When an ARP request is generated, the host that sends the request fills in the sender hardware address, sender IP address, and target IP address fields. The host that receives the request fills in the remaining field, target hardware address (that is, its physical address), exchanges the target IP and sender IP address fields, and sends back the message as an ARP reply (operation field equal to 2).

3.4.2　Reverse ARP (RARP)

This protocol is intimately related to the last one, given that it solves the opposite problem: a machine knows its physical address but not its IP address [9]. This happens, for example, in machines without a massive storage capacity, which need to use a *File Transfer Protocol* (FTP) type protocol over TCP/IP to load their set up program from a file server.

The function is very similar to ARP. The machine generates an RARP request message in broadcast mode, providing its physical address. Another machine, called the server, contains a table in its massive storage system with the assignation of addresses, and it replies with an RARP-reply type message. Note that this procedure will not be needed again until the machine is reinitialized.

3.5   Routing and Protocols

Routing in Internet is carried out at the IP level and concentrates on determining where IP datagrams are transmitted through the different physical networks making up the Internet before arriving at their destination. To study routing in the Internet, it is important to bear in mind that the architecture is made up of heterogeneous physical networks interconnected by gateways. A gateway is therefore a machine connected to more than one physical network and with routing capability (i.e., able to pass IP datagrams among the different networks to which it is connected).

Routing in the Internet is especially complex, due to the distinct characteristics of the component networks. Ideally, the routing must take into account aspects such as datagram length, network traffic, and the required type of service; however, normally it is based on a fixed table containing the shortest path to reach a specific destination.

Routing in the Internet is of two types: *direct,* where the datagram must be sent to a machine in the same physical network, and *indirect,* where it is necessary to access another physical network via a gateway.

3.5.1   Direct Routing

When a machine (host or gateway) wishes to send a datagram to another machine in the same physical network, the process is simple. The datagram is encapsulated in a frame, its physical address is determined from its IP address (either directly or using the ARP protocol), and it is sent using the network hardware.

The obvious question is how to know if the destination machine is within the same physical network. The answer is trivial. As has been seen, every IP address is made up of a prefix specifying the network and a suffix identifying each specific host. If the network prefixes identified by the origin and destination machines coincide, then they are in the same network.

3.5.2   Indirect Routing

When the network prefixes of the two addresses do not coincide, the destination machine is in a different network. In this case, routing is carried out (i.e., the most suitable gateway to direct the datagram to is chosen). This routing is done by the host and the intermediate gateways. When a gateway receives an encapsulated datagram through the physical network, it must examine the IP address of the destination and decide which gateway of another physical network to send it to, carrying out the suitable encapsulation for the new network.

To carry out indirect routing, the IP specifies an algorithm based on tables. This is discussed next.

3.5.2.1　The IP Routing Table Algorithm

The IP specifies a routing algorithm based on tables stored both in *hosts* and in *gateways*. An example of a routing table is shown in Figure 3.7.

It is obvious that a gateway only knows the next step toward the destination network and not the complete path. Bearing this in mind, and knowing that only the network prefix of each destination IP address is stored (the suffix specifying only the machine used during the final direct routing), the routing tables can be maintained with a reduced size. It should also be remembered that in this table a gateway or host only stores the gateways connected to its own physical network, which is always the first step toward any other destination network.

This routing mechanism has some drawbacks. For example, all of the datagrams directed to a specific network will follow the same path even though other alternative routes could be used at the same time. Moreover, the influence of the delays and the network throughput are not taken into account according to the type of traffic.

What happens when a destination network does not appear in the table? In this case, there is what is called a *default gateway*, which routes all of the datagrams whose destination network does not appear in the table. A default gateway is often the only entry in the routing tables of the hosts.

Finally, another important aspect of routing is how the tables are initialized and how they are updated as the network changes. In the beginnings of Internet, these tables were updated manually in the few gateways that existed, but this situation became impracticable. A series of protocols then developed as the Internet grew and changed its architecture permitting the initialization and updating of these tables automatically and dynamically through the interchange of routing information among the gateways.

0　　　　　　　　　　　　　　15 16　　　　　　　　　　　　　31	
IP destination network addresses	IP gateway address to reach the corresponding network
XX . XX . XX . 0	XX . XX . XX . XX
.....	.....

Figure 3.7　Example of routing table.

3.5.2.2 Gateway-to-Gateway (GGP)

This protocol was implemented in the Internet from its beginning [10]. The original architecture consisted of small LANs connected to the backbone of the Internet network, then called Arpanet.

The communication between this network and the LANs was carried out through a few gateways controlled directly by the Arpanet manager. Each gateway knew the routing information for all possible destinations and interchanged its tables with other neighboring gateways to maintain information coherency, using the GGP.

The GGP is based on the distance vector and the information exchanged consisted of pairs (*network address, distance from the gateway to this network*). The distance is measured in intermediate gateways (i.e., zero for a directly connected network with an increase of one for each intermediate gateway crossed in reaching the destination network). This distance vector–based method has the drawback that it does not consider the real capacity of each network. For example, a path through four Ethernet networks interconnected through three intermediate gateways can be shorter than a path through two networks connected by a low capacity serial line. However, it is possible to take this fact into account by artificially increasing the distance vector depending on its characteristics.

3.5.2.3 Routing Information Protocol (RIP)

Nowadays, the most commonly used protocol in the Internet is called the RIP, and it is a refinement of the GGP protocol based on the distance vector [11]. Two distinct types of machine participate in this protocol: active, if it can receive and transmit messages for updating tables, normally gateways, and passive, if it only receives messages, usually, hosts.

The distance vector in this type of protocol is based on the number of intermediate gateways, in the same way as in GGP. Nevertheless, the distance is increased artificially to reflect the real capacity of the different physical networks.

3.6 Mobility and the IP

In a book such as this one dealing with aspects relative to synergies between the IP world and wireless networks, it would be incomplete to deal with IP without discussing aspects related to mobility in the Internet.

Mobile personal computing devices are becoming ubiquitous as their prices drop and their capabilities increase. With the growing dependence of day-to-day computing on the distributed information base, providing network attachment to these devices is an essential requirement. Using wireless network interfaces, mobile devices can be connected to the Internet in the same way as desktop machines are connected using Ethernet or any other infrastructure. The major difference, however, is that mobile devices can move while in operation, which means that their point of attachment to the network can change from time to time. From a network's viewpoint, host movement constitutes a change in the network topology. It is natural that mobile users desire uninterrupted access to all networking services even while moving. However, host mobility introduces several new addressing and routing problems at the IP layer. The Internet routing system routes a datagram to a host based on the network number contained in the host's Internet address. If a host changes its point of attachment and moves to a new subnetwork, IP datagrams destined for it can no longer be delivered correctly.

It is useful to distinguish between two types of mobility in the Internet context:

- *Macro-mobility*, which implies the movement of a node among administrative domains. This implies a complete change of IP address and network settings (i.e., moving from your company in Europe to an airport in the United States).

- *Micro-mobility*, which is defined as mobility between access points within one administrative domain (i.e., one campus). IP addresses and administration are the same, however location and routing to mobile nodes is the problem.

3.6.1 Mobile IP

In this section the macro-mobility protocol known as *mobile IP* will be explained, which is discussed in depth in [12]. Mobile IP introduces the following new concepts:

- *Mobile node.* A node that can change its point of attachment to the Internet from one network to another.

- *Home address.* An IP address assigned to a mobile node within its home network. Its subnet prefix is the home subnet prefix.

- *COA.* An IP address associated with a mobile node while visiting a foreign network. The subnet prefix of this IP address is a foreign subnet prefix.

- *HA.* A gateway in a mobile node's home network with which the mobile node has registered its current COA. While the mobile node is away from home, the home agent intercepts any packet on the home network destined to the mobile node's home address, encapsulates it, and tunnels it to the mobile node's registered COA.

- *FA.* An FA is a gateway in a mobile node's visited network that provides routing services to the mobile node while registering. The FA detunnels and delivers datagrams to the mobile node that were tunneled by the mobile node's home agent. For datagrams sent by a mobile node, the FA may serve as a default gateway for registered mobile nodes.

Mobile IP solves the problem of mobility in Internet in the following way [13]. The FAs and HAs advertise their presence via agent-advertisement messages. A mobile node may optionally ask for an agent advertisement message from any local mobility agent by using an agent request message. When the mobile node receives an agent advertisement, it determines if it is in its home network or foreign network. If the mobile node detects that it is in its home network, it works without using the mobility services. On the contrary, if it detects that it has moved to a foreign network, it obtains a COA on the foreign network. The mobile node, operating away from home, then registers its new COA with its HA through the exchange of registration request and registration reply messages. The datagrams destined to the mobile node's home address are intercepted by its HA, tunneled by the HA to the mobile node's COA, received at the tunnel endpoint (either a foreign agent or at the mobile node itself), and finally delivered to the mobile node. Observe that in the communication in the opposite direction, the packets transmitted by the mobile node reach their destination without needing to be delivered previously to the HA.

3.6.2 Micro-Mobility Protocols

Several solutions have been proposed to solve the problem of micro-mobility in IP networks. Some of the best-known protocols are introduced here.

3.6.2.1 Cellular IP

Cellular IP is a micro-mobility protocol. It can be used to provide local mobility and handoff support [14]. It is therefore suitable for routing packets within a LAN, a campus network, or a *metropolitan area network* (MAN). In addition, it can interoperate with mobile IP to provide *wide area network* (WAN) mobility support.

Figure 3.8 shows one topology for a cellular IP network. It is made up of cellular nodes connected hierarchically. All cellular IP networks have a gateway at the boundary with the Internet backbone and a number of cellular nodes with an interface on a wireless link, known as *base stations*. Each cellular node always has only one uplink neighbor toward the gateway. The shortest path between a base station and the gateway is given by a chain of cellular nodes. This can be obtained starting from the base station and then considering one new node at a time, this being the uplink neighbor of the previous node until the gateway is reached.

When a mobile host wants to attach itself to the cellular IP network, it sends a signaling packet to the serving base station. This is then relayed hop by hop along the chain of uplink neighbors until it reaches the gateway. At each hop the packet creates a *soft state* (a soft state is a state that needs regular refreshing by special signaling packets or else it is cancelled within a definite period), which contains an association between the mobile host's home address and the downlink interface from which the message was received. The chain of these states corresponds with a path along which it is possible to route packets from the gateway to the mobile host. As a consequence, a packet bound for a mobile host in the cellular IP network must first reach the gateway and then follow the complete downlink path toward the mobile

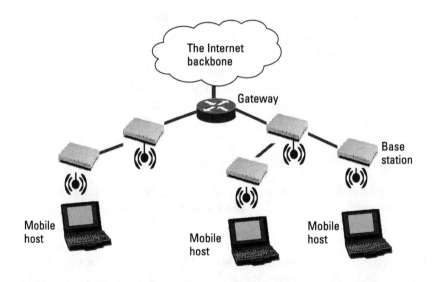

Figure 3.8 Example of the topology of a cellular IP network.

host, independently of where it was sent from, both from inside or outside the cellular IP network.

The states along the path must be regularly refreshed. This can be done by either signaling packets sent by the mobile host within the expiration time or data packets that flow along the path in the uplink direction. State refreshing through data packets enables a considerable reduction of signaling load in the whole network, saving processing capabilities.

Handoffs

Two options for handoffs are available, hard handoff and semi-soft handoff. When performing a hard handoff, a mobile host switches to the new base station and sends a signaling packet to it. This is then relayed hop by hop until it reaches the gateway in the same way as before and creates a new path. Packets that are traveling along the old path after the switching instant will be lost.

In the case of a semi-soft handoff, the mobile host waits for the new path to be created and continues receiving packets from the old base station before switching its receiver to the new one. The new path may partially overlap with the old one. In this case, the crossover node where the two paths join has a new state added for the new route created by a signaling packet coming from the mobile host; the old state pointing to the old route is left in place. As a result, there will be an interval of time in which the packets will flow through the new and the old path at the same time, to be finally relayed by both the old and the new base stations. During this time the mobile host can switch its receiver to the new base station and experience minimal packet loss. A subsequent signaling message will delete states that point to the old path.

When the new path is shorter than the old one, if no countermeasure is taken, some packets could be lost during the switching. This is because the packets that are relayed by the old base station will reach the mobile host delayed with respect to those forwarded by the new one. Consequently, when the mobile host makes the move to the new base station, it will not receive some of the packets it expects because they were transmitted by the new base station just a few instants before. The packets flowing along the new path are then kept and delayed for a definite time to compensate for this time gap and prevent the mobile host from losing packets when it switches to the new base station.

Idle Host Management

An important aspect of the cellular IP networks is that idle and active mobile hosts are managed separately to improve scalability and save battery. The databases used for routing packets are of two kinds: *route caches,* containing states only for those mobile hosts that are currently transmitting or receiving packets, and *paging caches,* present only in some nodes and containing both active and idle mobile hosts' states. Route caches are generally much smaller than paging caches. Consequently, they enable faster lookups that reduce node-crossing time.

Both route caches and paging caches have soft states. However, paging states have much longer expiration times, so they need to be refreshed less frequently. In addition, in the case of an idle mobile host, paging states do not have to be updated each time they move to a new cell. In fact, when a mobile host is idle, the corresponding states, which are stored in paging caches, are updated only if they move out of a group of cells called the *paging area.* This helps save battery in idle mobile hosts.

Interoperating with Mobile IP

Cellular IP can interwork with mobile IP to provide a larger mobility support. When a mobile host turns up in a cellular IP network, a preliminary registration with the local gateway has to be performed. During this phase, the host's registration request is submitted to an admission control module. The user is identified to find out its administrative permissions, and some technical and charging decisions can be taken. If the host passes this preliminary phase, mobile IP registration can begin.

If the mobile host is roaming, it has to register with the remote home agent in its home network. In this case, either the gateway or the mobile host can begin the mobile IP registration, using the gateway IP address as the mobile host's COA.

If a mobile host leaves a cellular IP network, its registration states will no longer be refreshed, and sooner or later they will be deleted. Alternatively, a mobile host leaving the network can explicitly erase its registration states by sending a tear-down message.

Columbia University, working in conjunction with Ericsson, has released an implementation of cellular IP to the public. Running on the FreeBSD operating system, this implements the gateway, base station, and mobile terminal functionality of cellular IP.

3.6.2.2 Handoff-Aware Wireless Access Internet Infrastructure (HAWAII)

HAWAII is another proposal dealing with micro-mobility [15]. It has also been developed with mobile IP in mind and has many common points with cellular IP. The approach is domain based, where a domain is defined by a hierarchy of gateways and base stations.

The edge gateway, connecting the access network to the Internet core, is called the *domain root gateway.* Location management is distributed, with many common points with cellular IP. Each gateway has a default route inside the domain pointing towards the domain root gateway. For every mobile node setting up its path, an entry for its IP address is added and associated with the appropriate interface.

HAWAII is transparent to mobile IP. A mobile node moving in a HAWAII-administered domain will not need to change its COA, and no communication with the home agent is required. Two special signaling packets are introduced, *power-up update* and *handoff update.*

In the handover case, two variants are proposed to deal with the different needs and functions, *forwarding* and *nonforwarding.* In the forwarding scheme, the packets from the old base station are forwarded to the new base station before they are diverted at the crossover gateway. In the nonforwarding scheme, data packets are diverted at the crossover gateway to the new base station, resulting in no forwarding of packets from the old base station. The forwarding scheme is optimized for access networks where the mobile host is able to listen/transmit to only one base station, whereas the nonforwarding one is geared towards networks where the mobile node can listen/transmit to more than one base station. The reason behind this separation is the desired minimum disruption and packet loss.

The signaling between the mobile node and the base station is performed by means of mobile IP signaling. The base station communicates in the fixed part of the HAWAII network with the other gateways and base stations using some special path setup packets that travel only within the HAWAII network. The consequence is that while gateways process only HAWAII messages, base stations must also implement mobile IP foreign agent functionality.

An example of HAWAII handoff functionality is presented in Figure 3.9. The mobile node communicates with the new base station, BS2, using mobile IP messages (arrow 1) and informs that BS1 was its previous base station. BS2 originates a path setup update (conveyed in a HAWAII handoff message, arrow 2) for BS1, so that future packets destined to the mobile hosts at BS1 are forwarded toward BS2. Then BS1 sends a HAWAII message to the gateway (arrow 3), which changes its forwarding entry from port B to port C. Finally, the gateway forwards the HAWAII message (arrow 4) to BS2, which updates its forwarding table and sends a mobile IP registration reply to the mobile host (arrow 5).

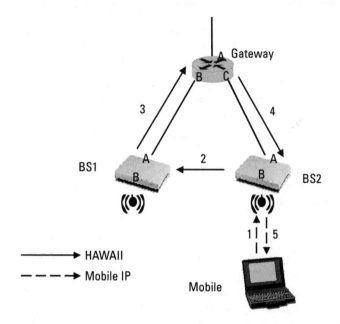

Figure 3.9   HAWAII example.

3.7   Transport Protocols

The transport layer is, without doubt, the heart of the TCP/IP protocol of Internet. Its main purpose, as in the *open system interconnection* (OSI) model, is to provide the application layer with point-to-point communications. This in turn enables the corresponding applications executed in different computers to transmit information through the Internet.

The basic task of the transport layer is to regulate the flow of information as well as ensure secure data transfer through the network, so it must implement error and sequencing control. For this, a system of *acknowledgement and nonacknowledgment* (ACK/NACK) with the other end of the communication is necessary, in such a way that packets that do not arrive correctly are retransmitted.

This layer will segment the flow of data arriving from the higher layer and pass the segments on to the IP layer. Furthermore, as the application layers of the two machines are normally made up of a set of programs running concurrently, it is likely that these establish simultaneous connections with the transport layer (transport multiplexing). Therefore, one of the basic tasks of this layer will be to add particular information identifying the origin and destination application.

3.7.1   User Datagram Protocol

Although it has just been explained that the main mission of the transport layer is to supply the application layer with secure connections with the application layers of other computers, so providing control services not provided by lower layers, this is not the whole truth. Within the existing set of protocols in the transport layer of Internet, there is one, the *User Datagram Protocol* (UDP) that provides no control, sequencing, or retransmission services [16]. This working scheme can be of interest in applications that do not require absolute reliability and at the same time cannot tolerate the delay introduced by the error control mechanisms based on retransmissions.

3.7.1.1   UDP Datagram Format

The UDP messages are denominated *UDP datagrams,* and they are made up of a header field and a data field, which arrive from the application layer.

The header transports the following information:

- The source port specifying the port of the application generating the datagram (normally set at zero unless the application requests a reply);

- The destination port specifying the port of the application to which the message is directed;

- Message length informing about the total length in octets of the message, including header;

- A check field whose use is optional and provides error control. It is obtained by dividing the whole message, including both the header and data, in blocks of 16 bits and then taking the one's complement of these blocks. The sum of one's complements of all of these blocks gives the result. If it is not used, it is set to zero. A virtual header is placed in front, which includes the source and destination IP addresses, the code of the transport protocol type, and the length of the message.

Figure 3.10 shows the format of a UDP datagram.

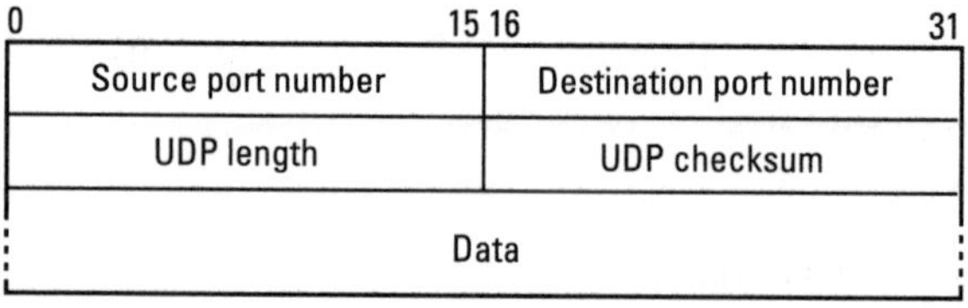

0	15 16	31
Source port number	Destination port number	
UDP length	UDP checksum	
Data		

Figure 3.10 UDP datagram format.

3.7.2 TCP

The TCP incorporates the features of robustness against losses and disordering of information necessary in a transport layer [17]. Its main mission is to handle and control the units of information of the transport layer, in TCP called *TCP segments,* which are sent to and received from the IP layer where, once inserted in the corresponding IP datagrams, they can be lost, duplicated, disordered, or corrupted as they travel through the Internet. In the same way, the TCP offers the application layer a service oriented to the connection, which, in the same way as with UDP, makes use of the port concept. Thus, a TCP connection is identified by a pair of endpoints, where each endpoint is made up of the pair (*IP host address, port number*).

3.7.2.1 TCP Segment Format

The TCP segment is made up of a header and a data field. Figure 3.11 shows the different fields making up this header, which are explained in the following paragraphs.

As can be seen and as was explained, each TCP segment contains the source and destination ports, with the aim of identifying the applications. The applications, along with the IP addresses, identify unequivocally each connection. The next two fields of the header, each of 4 bytes, permit the implementation of a reliable, connection-oriented data transmission service, identifying the last byte sent, sequence number field, and the first byte expected in the next segment, acknowledgment number field. The length of the header is measured in words of 32 bits, which is indicated in the HLen field. After 6 unused bits, reserved field, there are six 1-bit-long identifying flags conditioning the way each end deals with the segment:

- *URG* (urgent) indicates that the segment contains urgent data, and in this case, the urgent pointer points to the sequence number in the last octet in a sequence of urgent data.

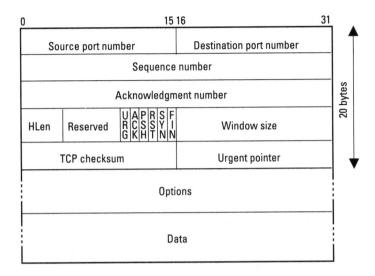

Figure 3.11 TCP segment format.

- *ACK* (acknowledgment) indicates that the acknowledgment number field is valid.

- *PSH* (push) indicates that the receiver must pass the data to the corresponding application as soon as possible without waiting to accumulate several segments.

- *RST* (reset) is used to indicate that the connection must be reset due to an error.

- *SYN* (synchronize) is used to synchronize the sequence numbers on setting up a connection.

- *FIN* (finish) indicates that data transfer has ended and the connection is to be closed.

The window size field indicates the quantity of bytes that can be accepted at each moment coming from the other end of the communication.

The checksum field is used to detect errors in the received segment. The algorithm used in TCP is the same as with IP, but in this case the checksum is done over the whole segment including the data.

The urgent pointer field indicates the end of these urgent data, given that the segment can also transport data that are not urgent.

The options field enables the inclusion of protocol extensions, some of which are shown later.

3.7.2.2   Establishing and Closing a TCP Connection

The establishment of the connection in TCP always uses a three-way hand-shake. To set up a connection, an entity sends a SYN segment, with the *sequence number* (SN) field with a value of *x*, which is the initial sequence number. The receiver responds with another SYN segment (*SN = y*), which confirms the reception of the previous segment, activates its ACK indicator, and puts the value *x + 1* in the *acknowledge number* (AN) field (next byte expected). Finally, the entity that sets up the connection sends another TCP segment confirming the previous reception, for which it activates the ACK indicator and puts the value *y* + 1 in the AN field. If the two ends generate SYN segments simultaneously, no error would be produced, given that both would be confirmed with the corresponding ACK, establishing the connec-tion. Figure 3.12 shows the procedure described.

The usual procedure for closing a connection is called *ordered closure,* in which an entity generates a FIN segment and awaits the corresponding ACK, when one direction of the connection closes (the other end could continue to send data). When the other entity decides to finalize the communication, the process is repeated and the connection is definitively closed.

3.7.2.3   Segment Transmission and Retransmission

In the normal exchange of information between two TCP entities, a sliding window scheme is followed, with confirmation from the receiver that it has

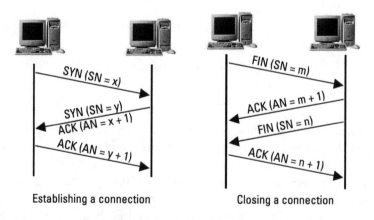

Establishing a connection   Closing a connection

Figure 3.12 Establishing and closing TCP connections.

received segments from the sender, with ACK segments. In the original implementation, the existence of negative acknowledgment is not contemplated, so the retransmission scheme is set up by expiration of the timers or by the reception of duplicate ACKs (after a modification in the initial recommendation, as will be discussed later).

Each time a packet is transmitted, a timer is set up that on expiration causes the retransmission of the segment. The duration of this depends on various aspects and is a crucial point in the behavior of TCP in wireless environments.

The retransmission process is affected by a mechanism of exponential backoff, in such a way that the TCP segment must be retransmitted repeatedly; the value of timeout is doubled each time until it reaches a defined maximum. For the first retransmission, however, the value of the *retransmission timeout* (RTO) is determined from the measurements of *round-trip time* (RTT), which are dynamically updating as the connection varies. In the original TCP specification, the RTT varied according to a smoothed estimator, *srtt* [18]:

$$srtt = \alpha \cdot R + (1 - \alpha) \cdot M \qquad (3.1)$$

In (3.1) α is the smoothed factor, R is the previous value of RTT, and M is the measurement that has just been made. In the initial implementation of TCP, the value of α was 0.9, although this parameter has evolved, and in the results shown in Chapter 5, the value of α has been chosen as 0.875.

Originally, the RTO was only evaluated from the estimation of RTT, multiplying it by a factor β with the recommended value of 2, although later it was demonstrated that it was more adequate to use the standard deviation of RTT to calculate the RTO. The value of the standard deviation, *mdev*, also evolved according to a smoothed estimation:

$$mdev = \delta \cdot mdev + (1 - \delta) \cdot M \qquad (3.2)$$

In this case, the value of δ used in the studies related to the TCP performance over LANs is 0.75. With both parameters, the value of RTO used is given by:

$$RTO = srtt + 4 \cdot mdev \qquad (3.3)$$

With retransmitted segments, an ambiguous condition can arise in the calculation of RTT, given that if a packet is retransmitted, on receiving the

ACK and taking the corresponding measurement to evaluate the RTT, it is not known whether it corresponds to the first transmission or to the retransmission. To avoid this ambiguity, the Karn algorithm is used [19], in which it is indicated that when retransmitting, the estimators of RTT cannot be updated until an acknowledgment is received for a segment that had not been retransmitted. Moreover, as there have been retransmissions and the backoff value has been applied to the RTO, this will be reused for the next retransmission. A new RTO will not be calculated until a segment that had not been retransmitted is acknowledged.

3.7.2.4 Congestion Control in TCP

In the conventional wired networks, the main problem suffered by the TCP connections was congestion, so a great effort was dedicated to implementing diverse procedures to reduce the loss in efficiency. These mechanisms have been added on to the different versions of TCP, giving rise to various implementations. The most important are analyzed next.

Slow Start and Congestion Avoidance

TCP assumes that all losses are caused by the congestion of intermediate gateways, which discard IP datagrams when they are near their saturation threshold. Several procedures have been included to enhance TCP's behavior, taking into account congestion situations. A TCP receiver imposes a flow control by informing the sender about the receive window, which indicates how much free buffer is available. Furthermore, a TCP sender controls its transmission rate by limiting the number of transmitted and nonacknowledged segments, using a variable known as congestion window (*cwnd*). When a new TCP connection is established, *cwnd* is set to one segment and is incremented by the number of segments acknowledged with every ACK (i.e., doubling its value each RTT). This phase is known as *slow start* [20]. Additionally, TCP ensures a reliable data delivery by retransmitting segments that are not acknowledged within some RTO interval. If it expires, TCP assumes that a segment has been lost and retransmits it; afterwards, the *cwnd* is set to one segment, and a variable known as *slow start threshold* (*ssthresh*) is then set to one-half of current window (the minimum of *cwnd* and the number of in-flight segments) and not less than two segments. Then, slow start is triggered again. When the value of *cwnd* reaches *ssthresh*, the increasing rate swaps from an exponential trend to a linear trend (one segment each RTT). This is known as the *congestion avoidance phase* [20]. Furthermore, if the same segment is lost consecutively, a backoff procedure is applied. In this way, the RTO is doubled after each retransmission.

It has been mentioned that a possible way of detecting a loss is a time-out expiration. RTO's value depends on the RTT measurements. Particularly, it takes into account its mean and its standard deviation using (3.3).

Fast Retransmit and Fast Recovery

TCP is required to generate an immediate acknowledgment (a duplicate ACK) that should not be delayed when an out-of-order segment is received. Its purpose is to notify the sender that a segment is being delivered out of order and to inform about what sequence number is to be expected. The sender does not know whether the reception of a duplicate ACK is caused by a lost segment or a reordering of segments, so it waits for a small number of duplicate ACKs (three acks) before performing a retransmission. This is the fast retransmit procedure.

The reception of duplicate ACKs is proof that just one segment has been lost, as the receiver can only generate them when another segment is received. Therefore, there is still data flowing between the two hosts, and it is not necessary to reduce the transmission rate greatly by invoking slow start. The fast recovery algorithm establishes that after a fast retransmit, congestion avoidance is executed instead of slow start.

The combination of the four algorithms that have been described previously gives rise to several TCP implementations, of which the two most important are described.

- *TCP Tahoe.* First implementation of TCP including a congestion control mechanism, specifically, slow start, congestion avoidance, and fast retransmit.
- *TCP Reno.* The fast recovery mechanism is added to TCP Tahoe, thus avoiding that the transmission path is emptied after executing fast retransmit, as would happen if slow start were executed instead of congestion avoidance.

3.7.2.5   TCP Options: Timestamp and TCP Selective Acknowledgment

In the format of the TCP header, as in Figure 3.11, it has been seen that there is a reserved space for the implementation of additional mechanisms in TCP. Thus, for example, one of the options that appears in most cases occurs in the establishment of a connection, given that when using the options field, the two participating entities negotiate what *maximum segment size* (MSS) will be used.

Timestamp

This option is added with the aim of updating the RTT more frequently to make it possible to adapt to the changes of transmission channel more quickly [21]. This option, as opposed to the usual RTT calculation, always updates its value when an acknowledgment arrives that increases the window, independently of whether it is a retransmitted segment. For its implementation, 10 bytes of the options field of the TCP header are used.

TCP Selective Acknowledgment

The ability to use this option is an important aspect, as it permits the *selective acknowledgment* (SACK) of segments [22]. In this way it avoids the possibility of retransmitting packets that had arrived correctly at the receiver, which causes an unnecessary use of bandwidth. The current implementation of TCP permits the acknowledgment of up to three nonadjacent blocks of segments.

3.7.2.6 Effects of the Wireless Links on the TCP

We have already mentioned that TCP presupposes that most losses produced in a connection are due to congestion. Therefore, it can be supposed that its behavior in channels where there is a relevant presence of errors, such as the radio channel case, will not be suitable. This leads to problems that greatly degrade the performance of protocols like TCP in networks with wireless links.

Three key independent factors affect the performance of reliable data transport in heterogeneous wireless networks:

1. The preponderance of packet losses due to wireless bit errors and user mobility;
2. Asymmetric effects and latency variation due to adverse interactions between media-access protocols and TCP;
3. Small transmission windows due to low channel bandwidths.

While TCP adapts well to network congestion, it does not adequately handle the vagaries of wireless media.

Wireless Bit Errors

TCP performance in many wireless networks suffers because of packet losses induced by wireless bit errors, which occur in bursts because of the nature of the wireless channel. Unfortunately, nowadays TCP wrongly attributes these losses to network congestion because of the implicit assumptions made by its congestion control algorithms. This causes the TCP sender to reduce its transmission window in response and often causes long timeouts during loss recovery, which keeps the connection idle for long periods of time. The result is degraded end-to-end performance. In addition, packet losses that occur due to user mobility cause the TCP sender to remain idle for long periods of time even after the handoff is completed, resulting in unacceptably low throughput [23].

Traditional approaches to addressing the bit-error problem either operate solely at the link-layer or split the end-to-end transport connection in two. Both of these approaches are fraught with problems. For example, pure link-layer protocols often interact adversely with TCP's loss recovery and timer mechanisms [24]. Split-connection protocols attempt to shield the sender from the wireless link by explicitly terminating the wired connection at the base station and using a separate transport connection over the wireless link [25] and thus not preserving end-to-end semantics.

An approach to solving this problem, called the *Berkeley Snoop Protocol* [26], exploits cross-layer protocol optimizations to improve performance.

Asymmetric Effects and Latency Variations

The second challenge arises due to the asymmetric nature of many wireless communication systems. An example of a network that exhibits asymmetry is a wireless cable modem network, where available bandwidth in the forward direction is much larger than that in the reverse path (a dial-up line or low-bandwidth wireless link). Such asymmetric bandwidth networks are becoming increasingly popular for economic reasons and predominantly asymmetric Web workloads.

Latency asymmetry occurs when latencies in opposite directions are different and variable, as in many packet-radio networks. A key problem observed in many multihop packet radio networks is the large and variable latencies that arise due to the reliable link-layer (for error control) and media access control protocols used in these networks. These variable latencies lead to ACK queuing at radio nodes and to highly variable round-trip times at the TCP sender. To ensure that packets currently in transit do not get retransmitted prematurely, TCP takes into account both the smoothed round-trip time estimate and the linear deviation in its timeout algorithm. Thus, timeouts in these networks lead to long idle periods of the order of several seconds.

Because protocols like TCP rely on timely ACKs for good performance, the capabilities and characteristics of the reverse path used for ACKs have the potential to seriously affect performance in asymmetric networks [27].

Low Channel Bandwidths

The bit rate commonly available between distant hosts in the Internet today is often limited and time varying. This is especially true in many wide-area wireless networks where maximum bandwidths (500 Kbps) are often orders of magnitude lower than their wired counterparts. In addition, today's dominant TCP application, the World Wide Web, uses short TCP connections because most Web transfers are short. For any connection, the average TCP transmission window size is roughly equal to the product of the available bandwidth and round-trip delay. This is because, in the steady state, the sender transmits a window's worth of data per round trip, and the ratio of these two is the available bandwidth. When the available bandwidth is low, so is the transmission window. When connections are short, the sender's window obviously does not grow beyond the length of the connection.

Small TCP transmission windows prevent the sender from recovering from losses without incurring expensive timeouts that keep the connection idle for long periods. As a result, end-to-end throughputs are significantly lower than the already-low maximum channel bandwidth: TCP transfers in these situations yield throughputs that are only a fraction of what is achievable. Thus, a key challenge is in enhancing TCP's loss recovery algorithms when low available bandwidths lead to small transmission windows.

References

[1]   Comer, D. E., *Internetworking with TCP/IP: Principles, Protocols and Architecture,* Upper Saddle River, NJ: Prentice Hall, 1991.

[2]   Siyan, K. S., *Inside TCP/IP,* Indianapolis, IN: New Riders Publishing, 1997.

[3]   Stevens, W. R., *TCP/IP Illustrated: The Protocols,* Reading, MA: Addison-Wesley, 1994.

[4]   Postel, J., "Internet Protocol," RFC 791, September 1981.

[5]   Braden, R., "Requirements for Internet Hosts—Communication Layers," RFC 1128, October 1989.

[6]   Almquist, P., "Type of Service in the Internet Protocol Suite," RFC 1349, July 1992.

[7]   Postel, J., "Internet Protocol," RFC 792, September 1981.

[8]   Plummer, D. C., "An Ethernet Address Resolution Protocol," RFC 826, November 1982.

[9]   Finlayson, R., et al., "A Reverse Address Resolution Protocol," RFC 903, June 1984.

[10] Hinden, R., et al., "The DARPA Internet Gateway," RFC 823, September 1982.

[11] Hedrick, C., "Routing Information Protocol," RFC 1058, June 1992.

[12] Perkins, C., "IP Mobility Support," RFC 2002, October 1996.

[13] Perkins, C. E., *Mobile IP*, Reading, MA: Addison-Wesley, 1998.

[14] Campbell, A. T., J. Gomez, and S. Kim, "Design, Implementation, and Evaluation of Cellular IP," *IEEE Personal Communications*, Vol. 7, No. 4, August 2000, pp. 42–49.

[15] Ramjee, R., et al., "IP-Based Access Network Infrastructure for Next-Generation Wireless Data Networks," *IEEE Personal Communications*, Vol. 7, No. 4, August 2000, pp. 34–41.

[16] Postel, J., "User Datagram Protocol," RFC 768, August 1980.

[17] Postel, J., "Transmission Control Protocol," RFC 793, September 1981.

[18] Jacobson, V., "Congestion Avoidance and Control," *Proc. of ACM SiGCOMM*, Standford, CA, August 1988, pp. 273–288.

[19] Karn, P., and C. Partridge, "Improving Round-Trip Time Estimates in Reliable Transport Protocols," *Proc. of ACM SiGCOMM*, Stowe, VT, August 1987, pp. 2–7.

[20] Jacobson, V., and M. Karels, "Congestion Avoidance and Control," *Computer Communication Review*, Vol. 18, No. 4, August 1988, pp. 314–329.

[21] Jacobson, V., R. Braden, and D. Borman, "TCP Extensions for High Performance," RFC 1323, May 1992.

[22] Mathis, M., et al., "TCP Selective Acknowledgment Options," RFC 2018, October 1996.

[23] Caceres, R., and L. Iftode, "Improving the Performance of Reliable Transport Protocols in Mobile Computing Environments," *IEEE Journal on Selected Areas in Communications*, Vol. 13, No. 5, June 1995, pp. 850–857.

[24] Balakrishnan, H., et al., "A Comparison of Mechanisms for Improving TCP Performance over Wireless Links," *IEEE/ACM Transactions on Networking*, Vol. 5, No. 6, December 1997, pp. 756–769.

[25] Bakre, A., and B. R. Badrinath, "I-TCP: Indirect TCP for Mobile Hosts," *Proc. 15th International Conf. on Distributed Computing Systems*, Vancouver, B.C., May 1995, pp. 136–143.

[26] Balakrishnan, H., S. Seshan, and R. H. Katz, "Improving Reliable Transport and Handoff Performance in Cellular Wireless Networks," *ACM Wireless Networks*, Vol. 1, No. 4, December 1995, pp. 469–481.

[27] Balakrishnan, H., V. N. Padmanabhan, and R. H. Katz, "The Effects of Asymmetry on TCP Performance," *Proc. ACM Mobicom Conf.*, Budapest, Hungary, September 1997, pp. 77–89.

4

Fundamentals of WLAN

4.1 Introduction

As has been described in previous chapters, almost every decade a new mobile system is developed and becomes a commercial service. Thus, the present-day 2G and emerging 3G mobile cellular networks are only some of the technologies moving toward mobile IP infrastructures. However, it seems certain that 3G alone will not be enough for a ubiquitous multimedia-capable IP infrastructure. Moreover, we believe that 4G is not likely to be a single air standardized and networking infrastructure like 3G. Obviously, it is still too early to define what 4G is, bearing in mind that the limits of 2.5G and 3G have still not been reached. Nevertheless, some guidelines can be laid down about what these systems may be like—the most obvious is that 4G is closely related to the concept of heterogeneous networks, including a large number of access networks, with the IP protocol stack as a common denominator, providing connectivity for all the users at any place and at any time.

Based on this scenario, it seems obvious that the WLAN infrastructures will play an important role in the immediate future as a complement to the existing or planned cellular networks. However, we would not be telling the whole truth if we suggested that WLAN was the only complement to cellular access networks. In this context, it should be mentioned that many cable operators, when faced with the excessive costs of *local multipoint distribution system* (LMDS) infrastructures, consider providing voice and data services to disperse rural zones, making use of WLAN equipment combined with low-cost *fixed wireless access* (FWA) equipment.

Reference [1] gives detailed information about WLANs. This chapter presents the main characteristics of three wireless systems: IEEE 802.11 in Section 4.2, HIPERLAN type 2 in Section 4.3, and MMAC in Section 4.4. These three wireless systems are defined in the corresponding standardization bodies in the United States, Europe, and Japan, respectively. Some practical issues of IEEE 802.11 deployment are given in Section 4.5.

4.2   The IEEE 802.11 Standard

In 1990, the IEEE formed a committee to develop a standard for wireless LANs, operating at 1 and 2 Mbps. For several reasons, but most importantly due to the existence in the market of different LAN products made by the corresponding manufacturers, the first of the standards took seven years to see the light of day. The IEEE 802.11 system [2] was approved in 1997, allowing work at data rates of 1 and 2 Mbps. In the fall of 1999, the standard was extended to break the 10-Mbps barrier [3]. Thus, the IEEE 802.11b was born, allowing reaching data rates of 5.5 Mbps and 11 Mbps. In parallel to this activity, a second group was working on a standard working in the 5-GHz band. This standard, known as IEEE 802.11a, allowed work at velocities of 6, 12, and 24 Mbps defining 9, 18, 36, and 54 Mbps as options [4].

Currently, the IEEE 802.11 standard body Task Group G is considering an even higher rate extension, for 11b networks, that will supply a payload rate in excess of 20 Mbps maintaining compatibility. Finally, the main standard-setting activities of the IEEE 802.11 committee involve enhancements to the MAC, 11e (quality of service), and 11i (security), along with the previously mentioned increase in velocity with respect to the existing standard, 11g.

4.2.1   IEEE 802.11 General Architecture

The IEEE 802.11 is a standard constituted by a PHY layer and a MAC layer. Over this layer, the standard foresees interfacing with the standard data LLC layer IEEE 802.2. The protocol architecture is depicted in Figure 4.1 where the PHY is chosen among three possibilities:

- *Frequency hopping* (FH) spread spectrum;
- *Direct sequence* (DS) spread spectrum;

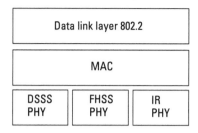

Figure 4.1 Protocol stack.

• *Infrared* (IR).

The system is constituted by the following entities:

• *Station (STA):* The object of the communication, in general a mobile station;

• *Access point (AP):* A special central traffic relay station that normally operates on a fixed channel and is stationary—can be partially seen as the coordinator within a group of STAs;

• *Portal (PO):* A particular access point that interconnects IEEE 802.11 WLANs and wired 802.x LANs. Thus, it provides the logical integration between both types of architectures.

Each of these entities implements the protocol structure of Figure 4.1 but employs different functions.

4.2.1.1 System Architecture

A set of STA, and eventually an AP, constitutes a *basic service set* (BSS), which is the basic block of the IEEE 802.11 WLAN.

The simplest BSS is constituted by two STAs that can communicate directly. This mode of operation is often referred to as an ad hoc network because this type of IEEE 802.11 WLAN is typically created and maintained as needed without prior administrative arrangement for specific purposes (such as transferring a file from one personal computer to another). This basic type of IEEE 802.11 WLAN is called *independent BSS* (IBSS).

The second type of BSS is an infrastructure BSS. Within an infrastructure BSS, an AP (which is a particular STA) acts as the coordinator of the BSS.

Instead of existing independently, two or more BSS can be connected together through some kind of backbone network that is called the *distribution system* (DS). The whole interconnected WLAN (some BSSs and a DS) is identified by the IEEE 802.11, as a single wireless network called *extended service set* (ESS). The whole scenario is shown in Figure 4.2.

The association between an STA and a particular BSS is dynamic, as discussed later; as a consequence, the set up of the system is automatic.

4.2.1.2   System Specification

Some basic features of the original IEEE 802.11 specification are sketched in Table 4.1.

It is important to remark that IEEE 802.11 does not impose any constraint on the DS (for example, it does not specify if the DS should be data link layer–based or network-layer based). Instead, IEEE 802.11 specifies a set of services that are associated with different parts of the architecture. Such services are divided into those assigned to STA, called *station service* (SS) and to the DS, called *distribution system service* (DSS). Both categories of services are used by the IEEE 802.11 MAC sublayer.

The services assigned to the STA are:

- Authentication/deauthentication;

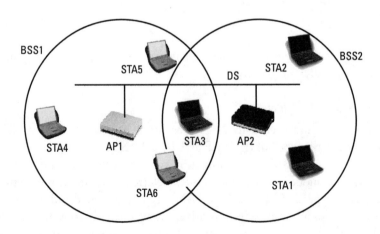

Figure 4.2   An ESS.

Table 4.1
IEEE 802.11 Original Specifications: Basic Features

Spectrum	2.4 GHz
Maximum physical rate	2 Mbps
Maximum data rate, layer 3	1.5 Mbps
MAC	CSMA/CA
Fixed network support	IEEE 802 wired LANs and others

- Privacy;
- *MAC service data unit* (MSDU) delivery to upper layer (IEEE 802.2 layer).

The services assigned to the DS are:

- Association/disassociation;
- Distribution;
- Integration;
- Reassociation.

The SSs are provided by all stations, including AP, conforming with IEEE 802.11, while the DSSs are provided by the DS and the most relevant will be explained in following sections.

These services are directly related to the IEEE 802.11 reference model, which is shown in Figure 4.3. When the description of the MAC sublayer is presented, the use of these services will be described.

Finally, and in relation to the IEEE 802.11 reference model, it is worth remarking that both MAC and PHY layers include two management entities: *MAC sublayer management entity* (MLME) and *PHY layer management entity* (PLME). These entities provide the layer management service interfaces through which layer management functions may be invoked.

4.2.1.3 Physical Layer

As depicted in Figure 4.3, the PHY is divided into two sublayers. The first layer is the *physical medium dependent sublayer* (PMD), which carries out the modulation and the encoding. The second layer is the *Physical Layer Convergence Protocol* (PLCP), which carries out PHY-specific functions, supporting common PHY SAP and providing a clear channel assessment signal.

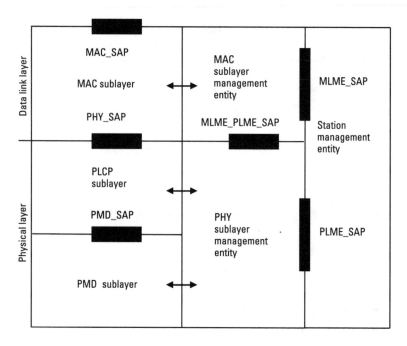

Figure 4.3　Protocol stack in detail. (*From:* [3]. © 1999 IEEE. All rights reserved. Reprinted with permission.)

This architecture has been designed to implement, under the same MAC, one PHY chosen among FH, DS, or IR, which have the characteristics described in Table 4.2.

For a more detailed description of aspects related to the physical layer, the reader might look at references such as [5, 6].

4.2.1.4　MAC Layer

The MAC layer is responsible for providing the following services:

- Asynchronous data service, which provides peer IEEE 802.2 entities with the ability to exchange MSDUs;

- Security services, which in IEEE 802.11 are provided by the authentication service and the *wired-equivalent privacy* (WEP) mechanism;

- MSDU ordering, whose sole effect, for the set of MSDUs received at the MAC service interface of any single station, is a change in the delivery order of broadcast and multicast MSDUs, relative to directed MSDUs, originating from a single source station address.

Table 4.2

PHY Specifications

	FH	DS	IR
Spectrum	2.4 GHz First channel at 2.402	2.4 GHz	Diffuse infrared (wavelength from 850 to 950 nm)
Subcarrier	1 MHz wide	11, 13, or 14 subchannels, each of 22 MHz	
Physical rate	1 and 2 Mbps	1, 2, 5.5, and 11 Mbps	1 and 2 Mbps
Modulation	2GFSK, 4GFSK	DBPSK, DQPSK, CCK (apply for IEEE 802.11b)	16 *pulse position modulation* (PPM) and 4 PPM
Other	Hop over 79 channels	11-chip Barker sequence	Nondirectional transmission

The general MAC frame format is shown in Figure 4.4 and consists of the following components:

- A MAC header, which comprises frame control, duration, address, and sequence control information;

- A variable length frame body, which contains information specific to the frame type;

- A *frame check sequence* (FCS), which contains an IEEE 32-bit CRC.

It is important to remark that there are four address fields in the MAC frame format. These fields are used to indicate the BSS identifier (BSS_ID), source address, destination address, transmitting station address, and receiving station address.

4.2.1.5 MAC Architecture

The MAC architecture can be described (see Figure 4.5) as providing the *point coordination function* (PCF) through the services of the *distributed coordination function* (DCF). The fundamental access method of the IEEE 802.11 MAC is a DCF known as *carrier-sense multiple access with collision avoidance* (CSMA/CA). The DCF will be implemented in all STAs, for use within both IBSS and infrastructure network configurations. The IEEE 802.11 MAC may also incorporate an optional access method called PCF, which is only usable on infrastructure network configurations. PCF exhibits

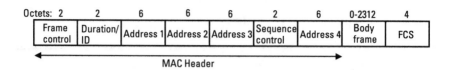

Octets: 2	2	6	6	6	2	6	0-2312	4
Frame control	Duration/ ID	Address 1	Address 2	Address 3	Sequence control	Address 4	Body frame	FCS

MAC Header

Figure 4.4 MAC frame format. (*From:* [3]. © 1999 IEEE. All rights reserved. Reprinted with permission.)

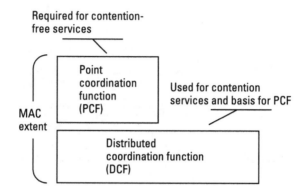

Figure 4.5 MAC architecture. (*From:* [3]. © 1999 IEEE. All rights reserved. Reprinted with permission.)

an extension of MAC functions and provides lower transfer delay variations to support time-bounded services.

Basic Access Method: DCF

The Basic Medium Access Protocol is a DCF that allows for medium sharing between STAs through the use of CSMA/CA. In this protocol, the STA, before transmitting, senses the medium. If the medium is free for a specified time, called *distributed interframe space* (DIFS), the STA executes the emission of its data. Otherwise, if the medium is busy because another STA is transmitting, the STA defers its transmission, and then it executes a backoff algorithm within a *contention window* (CW). This behavior of the CSMA/CA protocol is sketched in the Figure 4.6.

For reasons of efficiency, the backoff mechanism used in the DCF is discrete and the time following a DIFS is divided into temporal *slots*, whose duration depends on the physical medium used (fixed in such a way that a station can detect the transmission of a packet by any other station). The backoff procedure follows a binary exponential variation: for each transmission, the value of the interval is generated randomly, following a uniform

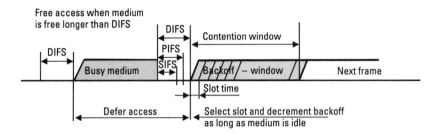

Figure 4.6 CSMA Protocol. (*From:* [3]. © 1999 IEEE. All rights reserved. Reprinted with permission.)

distribution in the range (0,CW). The CW depends on the number of times that a packet has been transmitted; for the first transmission, it takes a value defined as CW_{min} (minimum contention window), which is successively increased in whole powers of two, up to a maximum value of CW_{max}, as is shown in Figure 4.7. This represents an example in which CW_{min} is 7 and CW_{max} is 63. The values of these two parameters depend on the physical layer and are defined in the standard.

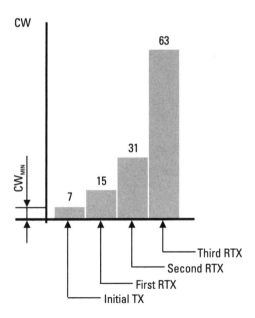

Figure 4.7 Example of the exponential increase of CW. (*From:* [3]. © 1999 IEEE. All rights reserved. Reprinted with permission.)

The exponential backoff algorithm is triggered whenever one of the following conditions occurs:

- When the STA senses that the medium is busy;
- After each transmission;
- After each retransmission.

Upon successful reception of a frame, the IEEE 802.11 MAC protocol requires that the destination generate an acknowledgment through an ACK frame, as is shown in Figure 4.8. Therefore, an ACK frame will be transmitted by the destination whenever it successfully receives a unicast frame of a type that requires acknowledgment, but not if it receives a broadcast or multicast frame of such a type.

The virtual carrier-sense mechanism is achieved by distributing reservation information announcing the impending use of the medium. The exchange of *request to send* (RTS) and *clear to send* (CTS) frames prior to the actual data frame is one means of distribution of this medium-reservation information. The RTS and CTS frames contain information, duration ID field, relative to the period that the medium is going to be reserved to transmit the actual data frame and the returning ACK frame. All STAs within the reception range of either the originating STA (which transmits the RTS) or the destination STA (which transmits the CTS) will learn of the medium reservation.

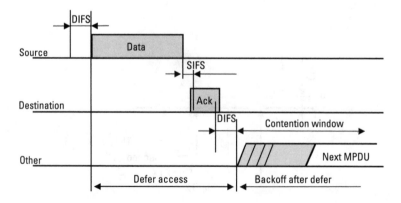

Figure 4.8 CSMA/CA + ACK. (*From:* [3]. © 1999 IEEE. All rights reserved. Reprinted with permission.)

Finally, this virtual carrier-sense mechanism makes use of the *network allocation vector* (NAV). The NAV maintains a prediction of future traffic on the medium based on duration information that is announced in RTS/CTS frames prior to the actual exchange of data. So STAs receiving a valid frame will update their NAV with the information received in the duration ID field. In this way, STAs reserve the medium and inform the PHY layer that the medium is busy. The same information is included within the ACK frame and is used when the data is fragmented. Figure 4.9 shows the whole scenario.

RTS/CTS frames are shorter than other frames, so that the mechanism reduces the overhead of collision. This is true if the data is longer than RTS/CTS; otherwise, it is not true. In the latter case, it is possible to transmit data without RTS/CTS transaction under the control (per station) of a parameter called RTS threshold.

In Figure 4.9, the *short interframe space* (SIFS) is one of the four possible *interframe spaces* (IFS), which are described as follows:

- SIFS is the shortest interval, employed to split transmissions belonging to a single dialogue (RTS-CTS or DATA-ACK). This value is fixed and is calculated in such a way that the transmitting station is able to switch back into the receive mode and is capable of decoding the incoming frame.

- *PCF IFS* (PIFS) is used by the AP to gain access to the medium before all other STA. Its value is calculated as SIFS plus a slot time.

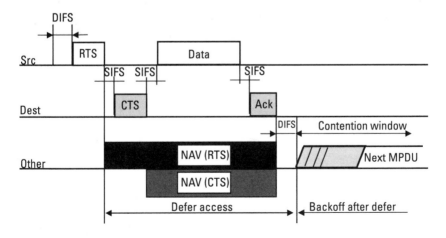

Figure 4.9 Medium reservation scenario. (*From:* [3]. © 1999 IEEE. All rights reserved. Reprinted with permission.)

- *Distributed IFS* (DIFS) is the time used by an STA having to start a new transmission. Its value is calculated as PIFS plus a slot time.

- *Extended IFS* (EIFS) is the longest IFS used by an STA, which receives a frame having an unknown duration that cannot be understood. This is to prevent collisions between STAs that do not hear RTS/CTS information.

Optional PCF

The PCF option is implemented to provide contention-free frame transfer and to thus support time-bounded services as well as transmission of asynchronous data, voice, or mixed. It is based on a *point coordinator* (PC), which has higher priority than other STAs. If it wants to transmit, it must wait a PIFS time, which is shorter than DIFS. Other STAs have to obey medium access rules of the PCF by setting their NAV at the beginning of a *contention-free period* (CFP). The operating characteristics of the PCF are such that all STAs are able to operate properly in the presence of a BSS in which a PC is operating, and, if associated with a point-coordinated BSS, are able to receive all frames sent under PCF control. It is also an option for an STA to be able to respond to a *contention-free poll* (CF-poll) received from a PC. An STA that is able to respond to CF-polls is referred to as to as being CF-pollable and may request to be polled by an active PC. CF-pollable STAs and the PC do not use RTS/CTS in the CPF. When polled by the PC, a CF-pollable STA may transmit only one *MAC protocol data unit* (MPDU), which can be to any destination (not just to the PC) and may piggyback the acknowledgment of a frame received from the PC using particular data frames. Figure 4.10 illustrates the described scenario.

It is important to note that the CFP is variable and ends with a CF-end frame transmitted by the AP.

4.2.1.6 Association Service

Before an STA is allowed to send a data message via an AP, it will first become associated with the AP. This service is needed either after an STA powers up or enters into a BSS area.

The STA needs to get synchronization information from the AP (or from other STAs, when in ad hoc mode) invoking the association service. For acquiring this synchronization information, the STA scans all channels by one of the following two strategies (depending on the value of the Scan-Mode parameter):

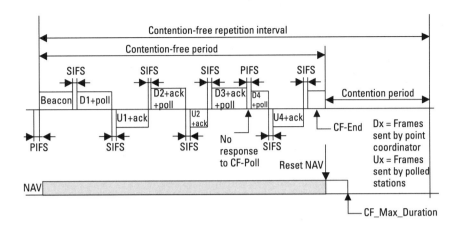

Figure 4.10 PCF access method. (*From:* [3]. © 1999 IEEE. All rights reserved. Reprinted with permission.)

- Passive scan, in which the STA will scan for beacon frames to collect synchronization information and to understand if the beacon frame comes from an infrastructure BSS or from an IBSS.

- Active scan, in which the STA transmits probe frames containing the desired *service set identifier* (SS_ID) and waits for a probe response from the BSSs within its area. The probe response is sent either from the AP of an infrastructure BSS or from the STA that generated the last beacon in an IBSS.

In general, beacon and probe frames contain information for joining a new network. Then STA chooses the BSS, which satisfies the desired SS_ID, sends an association request message (by setting the corresponding value in the control field of the MAC frame) to the selected BSS and waits for the corresponding association response frame. If there is not a BSS that satisfies its requests, STA may start an IBSS with its own characteristics. Figure 4.11 shows an example of active scanning in an ESS context.

Another two services can also be invoked: reassociation and disassociation. The reassociation can be invoked if an STA wants to move from one AP to another or if an STA wants to change association attributes while it remains associated to the same AP. The disassociation service is invoked whenever an existing association is to be terminated.

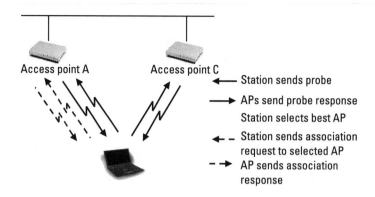

Figure 4.11 Example of association to a BSS.

4.2.1.7 Authentication and Privacy

IEEE 802.11 provides the ability to control LAN access via the authentication service. This is invoked by a STA to communicate its identity to other STAs of a BSS or IBSS.

There are two authentication techniques:

- *Open system authentication.* It is the default technique. A frame is transmitted by the STA invoking the authentication service, and a response frame is transmitted from the STAs involved. This constitutes the authentication result. If the result is successful, the STAs are mutually authenticated.

- *Shared key authentication.* It's the most secure technique. It consists of a set of transactions using a secret key exchanged through a secure channel different from those used by IEEE 802.11.

Shared key authentication requires the use of the WEP mechanism. The WEP algorithm is used by STAs to have confidential exchange of information. It employs a 40-bit secret key and only encrypts the payload of data frames. WEP uses the Rivest cipher (RC4) algorithm from RSA Data Security, Inc. [7].

4.2.1.8 Fragmentation

The process of partitioning a MSDU into smaller MAC-level frames, MPDUs, is called *fragmentation.* Fragmentation creates MPDUs smaller than the original MSDU length to increase reliability and so reducing the retransmission overhead in cases where channel characteristics limit reception reliability for longer frames.

The MPDUs resulting from the fragmentation of an MSDU are sent as independent transmissions, each of which is separately acknowledged (see Figure 4.12). The transmitting STA is not allowed to transmit a new fragment until one of the following happens:

- It receives an ACK.
- It decides that the fragment has been retransmitted too many times and then it drops the whole frame.

4.2.1.9 Synchronization

To synchronize other STAs within its BSS, an AP transmits a beacon frame where a time stamp is present; this prevents some drift of the STA synchronization. The beacon frame can be deferred only if a station is transmitting data.

4.2.1.10 Mobility

The STA can change the BSS where it is to be connected, using the active/passive scanning and reassociation service. In fact, while a STA is associated with a BSS, it can decide that the connection quality is poor, so it scans the medium to search for a more reliable connection. If the search is successful, it can decide to invoke the reassociation request to a new AP. If the reassociation response is successful, the STA has roamed to the new AP, which informs the DS. Normally, an old AP is notified through the DS. Otherwise, if the reassociation request fails, the STA tries to search for a new BSS.

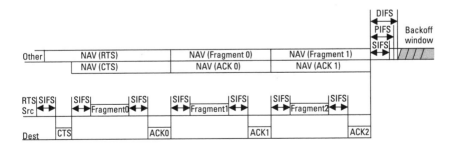

Figure 4.12 Transmission with fragmentation.

4.2.1.11 Power Saving

In an infrastructure network, the AP is the heart of the power management system [8]. If a STA wants to switch off the radio part for some period, it warns the AP through the frame control field of transmitted frames. In this case, the AP will buffer frames to STAs in *power saving* (PS) mode and then, during the transmission of the beacon frame, the AP broadcasts a *traffic indication map* (TIM) that contains an indication about the STAs that have frames buffered.

Thus, STAs listen for beacons after they wake up. If there are some buffered frames, they request the packets from the AP with a PS-poll frame and stay awake to receive data. If there are no buffered frames, the STAs revert to the sleeping mode. This is the case of unicast transmission.

On the other hand, for multicast/broadcast transmission, a special TIM called *delivery TIM* (DTIM) is a multiple of TIM period, after which the AP transmits multicast/broadcast packets. The PS mechanism is illustrated in Figure 4.13.

In ad hoc mode, every STA can transmit a beacon frame. After a beacon interval, every STA competes to transmit beacon frames with a backoff algorithm. The first STA wins and the others cancel their beacon and adjust their local time with the one included in the beacon frame that was transmitted.

The PS mechanism in ad hoc mode is similar to that described earlier, but in this case a special TIM is used, called *ad hoc TIM* (ATIM). ATIM is transmitted during the ATIM window, in which all of the STAs, including those operating in a PS mode, are awake. ATIMs are unicast frames that must be acknowledged by the receiver. After this, the receiver must be awake to wait for the announced packet. The mechanism is described in Figure 4.14.

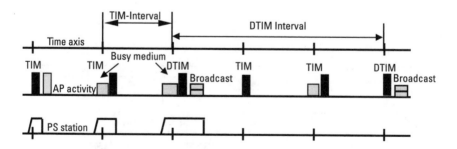

Figure 4.13   PS mechanism in an infrastructure BSS. (*From:* [3]. © 1999 IEEE. All rights reserved. Reprinted with permission.)

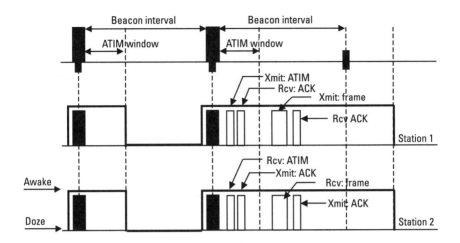

Figure 4.14 PS mechanism for an ad hoc network (DCF).

4.2.1.12 Multirate Support

Both in IEEE 802.11 standard and in its extensions (IEEE 802.11a and IEEE 802.11b), there is the possibility to transmit data at different bit rates. In order to support this option, a set of basic rates and operational rates has been defined. All control frames are transmitted at one of the basic rates, while data can be exchanged between two STAs at higher rates only if they support this.

4.3 HIPERLAN Type 2

4.3.1 Introduction

This section provides an overview of ETSI BRAN HIPERLAN standard, particularly highlighting type 2 [9]. A brief description of its system architecture and protocol reference model is provided, then focusing the attention on the MAC/*data link control* (DLC) layer.

4.3.2 HIPERLAN General Architecture

The BRAN family mainly comprises four different standards. We consider here, very briefly, the HIPERLAN type 1 standard, which provides a high-speed WLAN, and the HIPERLAN type 2 standard, which supplies short-range access to networks based on IP, UMTS, and ATM. Figure 4.15 shows the whole BRAN family.

The protocol architecture of HIPERLAN exhibits some differences between type 1 and 2. HIPERLAN type 1 employs an access method called *elimination yield–nonpreemptive priority multiple access* (EY-NPMA), which constitutes a kind of CSMA/CA that splits the procedure into three phases: priority resolution, elimination, and yield.

The protocol stack of ETSI BRAN HIPERLAN type 2 is constituted by three layers, each of them divided into user plane and control plane as shown by Figure 4.16. User plane includes functions related to transmission of traffic over the established user connections, while the control plane includes functions related to the control, establishment, release, and modification of the connections.

The "three" basic layers of HIPERLAN 2 are: PHY, DLC, and the *convergence layer* (CL), which is part of the DLC.

The PHY provides a basic data transport function by means of a baseband modem and an RF part. The transmission format on the PHY layer is a burst consisting of a preamble part and a data part. The modulation scheme chosen for the PHY layer is OFDM. OFDM was chosen due to its very good performance on highly dispersive channels.

The DLC layer consists of the *error control* (EC), the *radio link control* (RLC), and the MAC functions [8]. The DLC layer is divided into data transport and control functions. The user data transport part handles the data packets arriving from the higher layer via the *user service access point*

HIPERLAN type 1	HIPERLAN type 2	HIPER-ACCESS	HIPERLINK
WLAN	Wireless IP, ATM, and UMTS short range access	Wireless IP ATM remote access	Wireless broadband interconnect
MAC	DLC	DLC	DLC
PHY (5 GHz) 19 Mbps	PHY (5 GHz) 25 Mbps	PHY (various bands) 25 Mbps	PHY (17 GHz) 155 Mbps

Figure 4.15 The BRAN family.

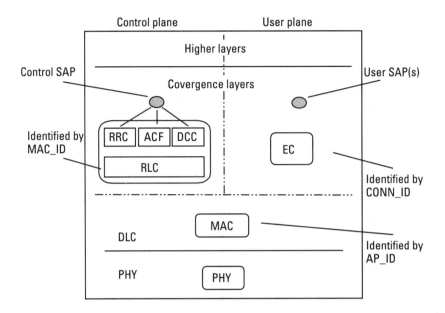

Figure 4.16 HIPERLAN 2 protocol reference model.

(U-SAP). The user data transport part also contains the EC, which performs an *automatic repeat request* (ARQ) protocol. The DLC protocol is connection oriented and for each DLC connection a separate EC instance is created. This allows different error control to be performed for different connections (e.g., depending on the service class). The control part contains the RLC function, which provides a transport service to the *DLC connection control* (DCC), the *radio resource control* (RRC), and the *association control function* (ACF).

Finally the CL is also divided into a data transport and a control part. The data transport part provides the adaptation of the user data to the DLC layer message format (DLC-SDU). If the higher layer network protocol is other than ATM, it also contains a *segmentation and reassembly function* (SAR) that converts higher layer packets (SDUs) with variable size into fixed size packets that are used within the DLC. The SAR function is an important part of CL because it makes possible the standardization and implementation of DLC and the PHY layers that are independent of the fixed network to which HIPERLAN-2 is attached. The control part of CL can use the control functions in the DLC (e.g., when negotiating CL parameters at association time).

4.3.3   System Architecture

The system is structured in a *centralized mode* (CM), even though a connection between two or more mobile stations is foreseen (see Figure 4.17). In fact, a direct link mode (DM) can be established between two or more mobile stations so that they can directly exchange information.

Two main entities are present in the centralized system:

- The *mobile terminal* (MT), which is the entity that wants to be connected to others and, if necessary, to external resources;
- The AP, the entity that coordinates the other MTs in its area and can control one or more sectors. Its protocol reference model differs from that of MTs for multiple MAC and RLC.

4.3.4   System Specification

Some basic features of the HIPERLAN-2 system are described in Table 4.3.

The integration of HIPERLAN into one of these fixed networks is due to the CL; this is constituted by a *common part* (CP), which also provides segmentation and reassembly according to the specific network layer employed, and a *service specific convergence sublayer* (SSCS).

In the following sections, the HIPERLAN-2's protocol reference model is examined in more detail.

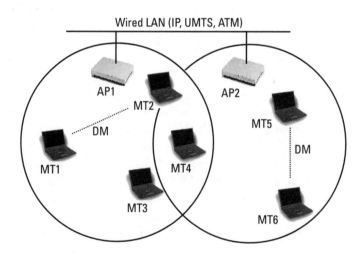

Figure 4.17 HIPERLAN-2 centralized architecture.

Table 4.3
Basic HIPERLAN-2 Features

Spectrum	5 GHz
Maximum physical rate	54 Mbps
Maximum data rate, layer 3	32 Mbps
MAC	Central resource control/TDMA/TDD
Fixed network support	IP/ATM/UMTS

4.3.5 Physical Layer

The basic transmission format on the PHY layer is a burst, constituted by preamble and data (where DLC-SDU is transmitted). As has been mentioned, HIPERLAN PHY layer is characterized by an OFDM, whose main features are summarized in the Table 4.4 [10]:

4.3.6 DLC Layer

The DLC layer represents the logical link between an AP and its associated MTs. The DLC layer implements a service policy that takes into account such factors as QoS characteristics of each connection, channel quality, number of terminal devices, and medium sharing with other access networks operating in the same area [11]. DLC operates on a per-connection basis, and its main objective is to maintain QoS on a virtual-circuit basis. Depending on the type of required service and the channel quality, capacity, and

Table 4.4
PHY Parameters

Spectrum	5 GHz
Channel spacing	20 MHz
Subcarrier per channel	52 (48 carry data and four are pilots that facilitate phase tracking)
Guard interval	Max 800 ns–Min 400 ns
Frequency selection	Single carrier with dynamic frequency selection
Forward error control	Convolutional code
Constrain length	Seven and generator polynomials (133, 171)
Modulation	BPSK, QPSK, 16QAM, 64QAM
PHY bit rate	From 6 Mbps to 54 Mbps

utilization, the DLC layer can implement a variety of means such as *forward error correction* (FEC), ARQ, and flow pacing to optimize the service provided and maintain QoS.

Two major concepts of the DLC layer are the *logical channels* and the *transport channels.*

A logical channel is a generic term for any distinct data path. A set of logical channel types is defined for different kinds of data transfer services offered by the DLC layer. Each type of logical channel is defined by the type of information it conveys and the interpretation of the values in the corresponding messages. Logical channels can be viewed as logical connections between logical entities, and so logical channels are mostly used when referring to the meaning of message contents. The names of the logical channel consist of four letters. HIPERLAN-2 DLC layer defines the following logical channels:

1. *Broadcast control channel (BCCH):* It conveys downlink broadcast control channel information concerning the whole radio cell.

2. *Frame control channel (FCCH):* Downlink, it describes the structure of the MAC frame. This structure is announced by *resource grant messages* (RGs).

3. *Random access feedback channel (RFCH):* Downlink, it informs the MTs that have used the RCH in the previous MAC frame about the result of their access attempts. It is transmitted once per MAC frame per sector.

4. *RLC broadcast channel (RBCH):* Downlink, it conveys broadcast control information concerning the whole radio cell. The information transmitted by RBCH is classified as:

   • Broadcast RLC messages;
   • Assignment of MAC_ID to a nonassociated MT;
   • Convergence layer ID information;
   • Encryption seed.

   RBCH is transmitted only when necessary.

5. *Dedicated control channel (DCCH):* It transports RLC messages in the uplink direction. A DCCH is implicitly established during association of an MT.

6. *User broadcast channel (UBCH):* Downlink, it transmits user broadcast data from the CL. The UBCH transmits in repetition or unacknowledged mode and can be associated or unassociated to LCCHs.

7. *User multicast channel (UMCH):* Downlink, it is employed to transmit user point-to-multipoint user data. The UMCH is transmitted in unacknowledged mode.

8. *User data channel (UDCH):* Bidirectional, it is employed to exchange data between APs and MTs in CM or between MTs in DM. The UDCH is associated or not to LCCHs.

9. *Link control channel (LCCH):* Bidirectional, it is employed to exchange ARQ feedback and discard messages both in CM and in DM. The LCCH is also used to transmit *resource request messages* (RRs) in the uplink direction (only in CM) and discard messages for a UBCH using repetition mode. LCCHs may or may not be associated with UDCHs/UBCHs.

10. *Association control channel (ASCH):* Uplink, in this case the MTs that are not associated to an AP transmit new association and handover requests.

The logical channels are mapped onto different transport channels. The transport channels provide the basic elements for constructing *protocol data units* (PDUs) and describe the format of the various messages (e.g., length, value representation). The message contents and their interpretation, however, are subject to the logical channels. The transport channels are named and referred to with three-letter abbreviations. The following transport channels are defined in the DLC layer:

1. *Broadcast channel (BCH):* Downlink, it contains 15 bytes of radio cell information such as identification of the AP and its current transmitted power.

2. *Frame channel (FCH):* Downlink, its length is a multiple of 27 octets. It contains a description of the way resources have been allocated and can also contain an indication of the empty parts of a frame.

3. *Access feedback channel (ACH):* Downlink, its length is 9 octets. It contains information on access attempts made in the previous RCH.

4. *Long transport channel (LCH):* Downlink and uplink, its length is 54 octets. It is used to transmit DLC user PDUs (U-PDUs of 54 bytes with 48 bytes of payload).

5. *Short transport channel (SCH):* Downlink and uplink, its length is 9 octets. It is used to exchange DLC control PDUs (C-PDU of 9 bytes).

6. *Random channel (RCH):* Uplink, its length is 9 octets. It is used for sending control info when no granted SCH is available. It carries RRs as well as ASCH and DCCH data.

Figures 4.18–4.20 show the mapping of logical channels into transport channels in centralized mode and in direct mode.

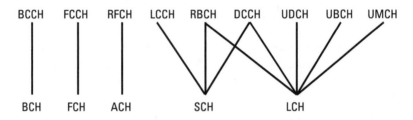

Figure 4.18 Mapping between logical and transport channels for the downlink.

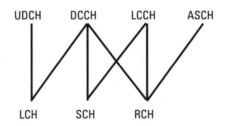

Figure 4.19 Mapping between logical and transport channels for the uplink.

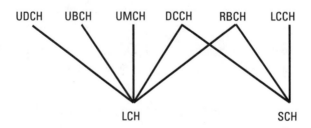

Figure 4.20 Mapping between logical and transport channels for the direct link.

4.3.6.1 MAC Layer

MAC protocol is based on TDMA/TDD, and frames exhibit a repetition period of 2 ms. The APs control the allocation of resources and generally determine if two MTs can directly exchange information. An AP needs to know the state of its own buffers and of the buffers in the MTs, and the allocation of resources are conveyed by RG. The MTs may request resources, in terms of transmission capacity, using RRs where there is an indication of the state of AP buffers. Optionally, MTs can request fixed capacity allocation over multiple frames.

4.3.6.2 MAC Operations

The MAC protocol involves the following elements:

- A scheduler, centralized in the AP, which determines the composition of the MAC frame. It respects the rules of each transport channel that it manages. In order to compose the frame, it employs the information included in RRs transmitted by MTs and the state of its own downlink transmission buffers.

- A process in APs and in MTs that receives and transmits PDUs in accordance with the MAC frame defined by the AP.

- A process that maps logical channels onto transport channels.

- MAC entities that exchange control information such as that in FCCH and resource request or feedback for the contention channel.

The MAC protocol provides an indication of particular errors in uplink RRs and discarding PDUs. This indication is included in two-bit fields; the first one (called *error reason bits*) contains the summary of errors in either the BCHs and FCHs or LCHs. The second (called *channel quality bit*) contains the overall channel quality.

AP MAC Operation

One of the operations carried out from APs is the calculation of the frame composition; the BCH, FCH, and ACH are also prepared and transmitted. They have the ability to transmit (in the downlink phase), receive, and process (in the uplink phase) the PDUs from and to the MTs, according to the current frame composition and rules. This will be performed both in CM and DM (in fact the AP can also be involved in direct link operations).

Finally, the APs receive and process the PDUs transmitted by the MTs in the RCH and prepare the corresponding ACH.

MT MAC Operation

The MTs receive and process the BCH and the FCH, and have the ability to evaluate the structure of the current frame. They also have the ability to transmit (in the downlink phase), receive, and process (in the uplink phase) the PDUs from and to MTs according to the current frame composition and rules. These are fulfilled both in CM and DM. Finally, they can access to RCH and, in the following frame, evaluate the ACH.

4.3.6.3 MAC Frame

The MAC frame structure is shown in Figure 4.21. Whenever two or more STAs are involved in the direct mode, their data have to be transmitted into the *direct link* (DiL) phase, which will be present. We underline that the duration of the BCH is fixed, while the others are dynamically adapted by AP depending on the traffic situation.

4.3.6.4 MAC Addressing

Each MT has an associated MAC_ID, which is unique for an AP and is assigned at the moment of association. The MAC_ID is coded with 8 bits, where the values 0 and 254–255 are reserved for special purposes.

Each connection, in the same MT, is addressed with a DLC connection ID (DLCC_ID), which is coded with 6 bits. In the centralized mode, a DLCC_ID and the MAC_IDs of AP and MT identify their communication. On the other hand, in direct mode, both the DLCC_ID and the MAC_IDs of MTs involved in a communication identify their connection. The *network*

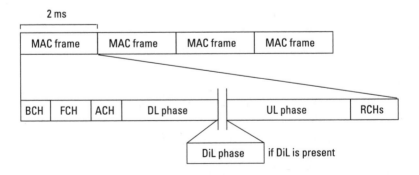

Figure 4.21 Basic MAC frame structure (DiL optional).

identifier (NET_ID) identifies APs, which belong to the same network of a certain operator. Finally, each AP has an AP_ID coded with 10 bits.

4.3.6.5 Access to RCH

Each MT maintains a contention window, CW_a, derived from a number a, which is the number of retransmissions made by the MT. For the first transmission, a is equal to 0. CW_a controls the access to the RCH, and its size is calculated as follows:

1. Initial attempt: $a = 0$, $CW_0 = n$
2. Retransmission: $a \geq 1$, $CW_a = 256$ if $2^a \geq 256$; 2^a if $n < 2^a \leq 256$ or n if $n \geq 2^a$

CW_a is constituted by $\max(2^a, n)$ RCHs, and each of them are numbered in ascending mode from 1 to CW_a size. According to its retransmission situation, each MT randomly chooses a number r between 1 and CW_a and starts counting r RCHs. The MT can only access the rth RCH. Finally, if it receives the ACH with a positive feedback, it then resets a to 0.

4.3.7 Other DLC Entities

Most of DLC services are executed by RLC and EC entities. In particular, the RLC implements the control plane of the DLC layer [12]. It includes three entities, each one specified for different functionality. These associated signaling entities are the *DLC connection control* (DCC), the RRC, and the ACF.

The DLC connection control is specialized in the appropriate signaling procedures needed to establish or release a connection. The connection setup function starts with a request, which is mainly initiated by the MT. During this action, the connection characteristics are being negotiated. If the AP accepts the MT's requirements, an acknowledgment is sent back. The DCC entity also supports release-signaling procedures and capabilities of modification of established connections.

The ACF supports all of the functions related to the exchange of information about link capabilities and association of the MT with the corresponding AP. If the MT finds the most suitable AP to associate with (this decision is based on the signal measurements performed by the MT), it will then request a MAC ID from the AP. The procedure continues with information exchange about the supported PHY modes, the convergence layer, and the authentication and encryption procedure. Encryption starts with a key exchange in order to guarantee security between sessions. HIPERLAN-2 supports both *data encryption standard* (DES) and 3-DES encryption algorithms. For the authentication procedure, the message digest (MD5), *hash-based message authentication code* (HMAC), and the *Rivest, Shamir, Adleman* (RSA) algorithms are supported. After the association is completed, the MT requests one or more DLC user connections. Disassociation may be done in two ways, explicitly or implicitly. The former is MT initiated and occurs when the MT has no communication demands from the network, while the latter is a special situation, which happens after a long period of sleep mode from the MTs side.

The radio resource control involves four main functions: the handover, the *dynamic frequency selection* (DFS), the MT alive, and the power-save procedures.

The handover is mainly MT initiated. It requires measurements of the link's quality from the MT side in order to decide the handover action (handover procedure is described in detail in Section 4.3.8).

The dynamic frequency selection is the procedure that automatically assigns frequencies to each AP for communication. This procedure takes into account the interference issues by collecting results from the APs and their associated MT measurements.

The MT alive function provides the AP with the capability to figure out if any associated MT is not transmitting. A timer may be set in order to limit the sleep mode of MTs. If no response from the MTs arrives to the AP, a disassociation procedure begins.

The power-save procedure is used to define the appropriate signaling for transmitter power control and definition of sleeping mode of MTs.

The HIPERLAN-2 error control entity supports three different modes of operation:

- Acknowledged mode;
- Repetition mode;
- Unacknowledged mode.

Acknowledged mode provides with reliable transmissions using retransmissions to compensate for the poor link quality. The retransmissions are based on acknowledgments from the receiver. The ARQ protocol that is used is *selective repeat* (SR), and EC allows for various transmission window sizes to be used depending on the requirements of each connection. In order to support QoS for delay-critical applications (e.g., voice, real-time video), the EC may also utilize a discard mechanism for discarding LCHs that have exceeded their lifetime. Figure 4.22 illustrates data and control flow in acknowledged mode.

Repetition mode provides a reliable transmission by repeating the LCHs. In repetition mode, the transmitter transmits new LCHs in consecutive order and is allowed to make arbitrary repetitions of each LCH. No feedback is provided by the receiver. Repetition mode is used for the transmission of UBCH. Figure 4.23 illustrates data and control flow in repetition mode.

Finally, unacknowledged mode provides an unreliable, low-latency transmission. In unacknowledged mode the data flows only from the transmitter to the receiver. No ARQ retransmission control or discard messages are supported. Unacknowledged mode is used for the transmission of UMCH, DCCH on LCH, and RBCH on LCH but can also be used for UDCH (UDCHs for a certain connection can be sent in either acknowledged or unacknowledged mode). Figure 4.24 illustrates data and control flow in unacknowledged mode.

4.3.8 Handover Issues

The handover capabilities supported for HIPERLAN-2 so far are mainly MT initiated. However, there is an AP-initiated capability for handover in case an AP wants to decrease its load in order to increase its performance or for other reasons. This action should not be executed if the MT is not capable of a handover procedure. The MT handover can be performed in three different ways: *sector handover, radio handover,* and *network handover.*

Sector handover is the procedure that takes place when an MT moves from one sector to another. This implies that the MT works in a sectorized cell. The MT requests a handover via the old sector. If the communication between the old sector and the MT is feasible, the MT will change to the new sector; otherwise, it will have to send a request to the new sector and then

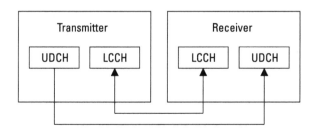

Figure 4.22 Data and control flow in acknowledged mode.

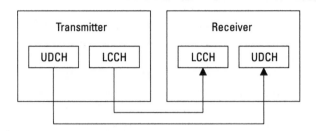

Figure 4.23 Illustration of the data and control flow in repetition mode.

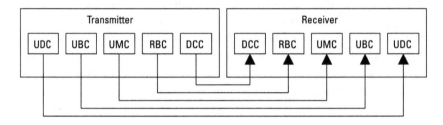

Figure 4.24 Data and control flow in unacknowledged mode.

change to the new sector. In both cases the AP should reply with an ACK message.

Radio (intra-AP) handover requires a multitransceiver environment per cell. It may occur when an associated MT moves from the coverage area of one *AP transceiver* (APT) to the coverage area of another, when both belong to the same AP. Again the whole procedure is MT initiated and the MT notifies the AP about the handover action. In case the MT returns to its old area, it still must notify the associated AP. Communication between these two entities is implemented via the old APT, until the AP receives the handover notification message. Except for notification, the MT must send a request in order to inform the target AP. The handover procedure will not end until the MT receives a radio handover complete message.

Network (inter-AP) handover is the most complicated because it involves higher layer functionality as well. It occurs during the movement of an associated MT from one AP to another. This kind of handover basically supports the same signaling procedure as the radio handover but also includes a security mechanism for justifying that the MT is really the one that made a network handover from the old AP to the new AP. This type of handover may require a signaling protocol specification at higher layers, in order to maintain the association with these and to guarantee a sufficient usage of the established connections. A radio and network handover scenario can be seen in Figure 4.25.

4.3.9 CL

The CL has two main functions. First, it adapts the higher layer service requirements to the DLC layer capabilities, and second, it modifies the traffic in such units (packets) so that the latter can be accepted to the layers above and below it. There are two types of CL defined so far. One is cell based and intended to be used for interconnection with the ATM networks, while the other is packet based and specified for fixed network use. The convergence layer is divided in two main parts: the CP and the SSCS. The common part's purpose is to segment and reassemble the packets that pass through it and to add some padding bits in order to build compatibility with the other layers' packet formats. Each service-specific convergence sublayer

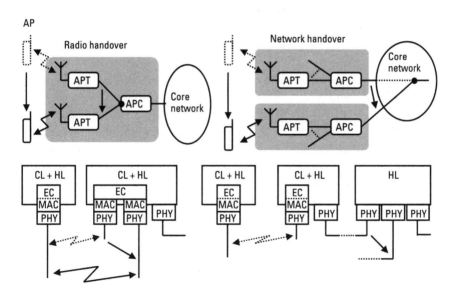

APT: Access point transceiver
APC: Access point controller
HL: Higher layers

Figure 4.25 Radio and network handover procedure.

adapts traffic for interworking with a fixed network. So far, only an Ethernet interface is specified.

4.3.10 QoS Support in HIPERLAN-2

A very important issue, in network implementations, is for the customer to be able to use the network for every type of traffic needed. However, traffic needs grow more and more along with user demands for better services. Different kinds of traffic flows require different treatment, based on their bandwidth, delay, jitter, or bit error rate characteristics, in order to avoid incompliant services. To handle this problem, HIPERLAN-2 specifications involve the following aspects.

HIPERLAN-2 supports QoS in different ways. This can be focused on three major points of the current HIPERLAN-2 architecture.

First, the connection-oriented nature of the HIPERLAN-2 makes it straightforward to implement support for different per-connection QoS. Data is transmitted between the AP and the MT only after a connection setup is completed. This procedure requires signaling functions, which are related to the DLC user connection control part of the RLC. Most supported connections are bidirectional and point to point. Multicasting is also supported in unidirectional mode (from AP to MT).

The main feature of QoS support is the mapping onto the RLC PDUs. This takes place during the connection setup procedure, when each characteristic of the connection is being negotiated between the AP and the MT. The DLC provides two kinds of setup: an AP-initiated *DLC user connection* (DUC) setup procedure and a MT-initiated DUC procedure. In both procedures, the AP decides whether to finally establish a connection. Mapping of the QoS demands can be easily implemented inside the data format of the connection setup signal. The number of attributes negotiated is currently limited by the size of the LCH.

The last point of interest concerning QoS in the HIPERLAN-2 implementation is the different capabilities of the currently specified SSCS. Assuming the Ethernet SSCS to be the only existing standard so far that can be used below IPv6, its capabilities can be extensively used to provide additional QoS performance. This can be achieved using the IEEE 802.1p priority based QoS scheme, which is supported by Ethernet SSCS. According to this standard, eight different priority levels are defined. These priorities are mapped to queues, and the priority information is carried in a new tag header inserted in the IEEE 802.3 frame. This header helps to distinguish each traffic source type, as each type of traffic is being assigned a specific number. The use of the IEEE 802.1p standard is optional for the AP and MT, while the normal supported standard is the best-effort scheme.

4.4 MMAC-PC

MMAC-PC stands for Multimedia Mobile Access Communication Systems Promotion Council. The main purpose of the MMAC systems is the provision of ultra-high-speed transmission of high-quality multimedia information at any time and anywhere with seamless connections to optical-fiber networks. The position of MMAC is shown in Figure 4.26.

Four segments are covered by this standard:

- *High-speed wireless access (outdoor and indoor)*. These are mobile communications systems that can transmit at up to 30 Mbps using the 25-/40-/60-GHz bands and with a bandwidth from 500 to 1,000 MHz. The typical service areas are public spaces (both outdoor and

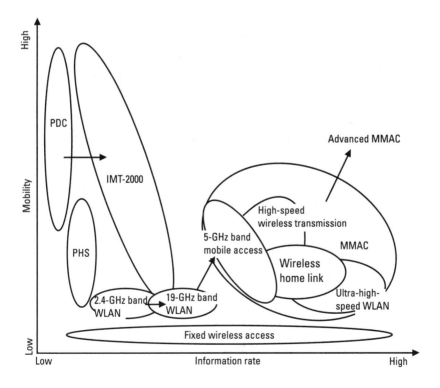

Figure 4.26 The MMAC position.

indoor) and private spaces (indoor premises). Only pedestrian mobility is supported (with handover), and the targeted terminals are notebook-type PCs and similar devices.

- *Ultra-high-speed WLAN (indoor).* This is a WLAN that can transmit at up to 156 Mbps using the 60-GHz band. The allocated bandwidth is between 1 and 2 GHz. It can be used for high-quality TV conferences, so desktop PCs and workstations are the potential terminals.

- *5-GHz-band mobile access (outdoor and indoor).* This is an ATM-type wireless access and Ethernet-type WLAN using the 5-GHz band. Each system can transmit at up to 25 Mbps for multimedia information. The typical service areas are public spaces (outdoor and indoor) and private space (indoor premises). Low mobility and roaming will be supported. Terminals like notebook-type PCs and handy terminals are applicable for this specification. In principle, IEEE 802.11a provides the general framework for physical and MAC layers.

- *Wireless home link.* Wireless home link can transmit at up to 100 Mbps using the 5-/25-/40-/60-GHz frequencies allocating a bandwidth greater than 100 MHz. Its main use is foreseen for transmitting audio-visual information between PCs and audio equipment. Low mobility will be supported, including roaming.

4.5 Deployment of the IEEE 802.11 Infrastructure—Some Practical Considerations

This last section of the chapter attempts to provide some considerations to bear in mind when planning the deployment of an IEEE 802.11 infrastructure for indoor and outdoor environments. For the latter case, we also describe the equipment provided by a manufacturer, Avaya, that permits a true implementation of wireless IP in outdoor environments.

Given that until now, in the market a setup compliant with the IEEE 802.11b standard is the norm, the following description is based on that specification.

4.5.1 The ISM Band and Channel Allocation

The working band assigned to IEEE 802.11b corresponds to an *industrial, scientific, and medical* (ISM) band centered on 2.4 GHz. This band is regulated in a different way in Europe, Japan, and the United States. Even at a European level, France has a particular assignation of this band that motivates its separate inclusion in Table 4.5, which shows the set of channels defined for operation with an infrastructure of the IEEE 802.11b type working in DS spread spectrum mode.

In Table 4.5, the frequencies indicated correspond to the central frequency of each channel. Given that the bandwidth of each channel is 22 MHz, these channels overlap in part of the band assigned to each one. Figure 4.27 shows this situation for the U.S. case.

Most equipment available in the market permits the choice of one channel or another through a configuration menu provided by the manufacturers for this purpose. Thus, Figure 4.28 shows the configuration menu of an Avaya access point that permits this selection.

Due to the overlap of the part of the frequency band that can be produced among the different channels when more than one WLAN infrastructure is operating in neighboring environments, it is necessary to maintain a minimum separation among the channels to be used. The limit case is produced when an AP has the capacity to place two PCMCIA cards with the

Table 4.5
Channel Sets for the IEEE 802.11b DS Spread Spectrum (in GHz)

	Europe	France	Japan	U.S.
1	2,412	—	2,412	2,412
2	2,417	—	2,417	2,417
3	2,422	—	2,422	2,422
4	2,427	—	2,427	2,427
5	2,432	—	2,432	2,432
6	2,437	—	2,437	2,437
7	2,442	—	2,442	2,442
8	2,447	—	2,447	2,447
9	2,452	—	2,452	2,452
10	2,457	2,457	2,457	2,457
11	2,462	2,462	2,462	2,462
12	2,467	2,467	2,467	—
13	2,472	2,472	2,472	—
14	—	—	2,484	—

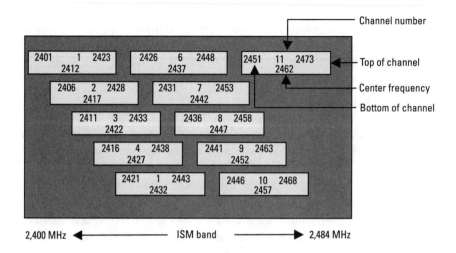

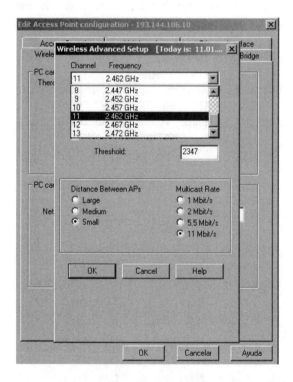

Figure 4.27 Channel allocation in the United States for the 2.4-GHz ISM band. (Courtesy of Avaya.)

Figure 4.28 Selection of the channel and frequency through a configuration menu. (Courtesy of Avaya.)

aim of having, for example, two WLAN infrastructures working independently. In this case, Figure 4.29, obtained from the specifications provided by Avaya, shows the combinations among channels that enable the cohabitation of the wireless segments.

4.5.2 Signal, Interference, and Radio Coverage

When a WLAN infrastructure is to be deployed, with the aim of guaranteeing an adequate function of the equipment while saving effort, it is beneficial to carry out a previous study relating to the points in which coverage is envisioned. For this, it is necessary to know aspects relating to the transmitted power, sensitivity of the equipment, possible interference sources, and the propagation environment. In this section, some practical considerations are provided that allow the reader to approach the deployment of a wireless infrastructure similar to those dealt with in the chapter.

4.5.3 Signal and Interference in the ISM Band

The ISM band, as happens with the rest of frequency bands, has some established maximum levels of output power with the aim of minimizing

Adapter 1's	Adapter 2's channel #										
channel #	1	2	3	4	5	6	7	8	9	10	11
1									x	x	x
2										x	x
3											x
4											
5											
6											
7											
8											
9	x										
10	x	x									
11	x	x	x								
12	x	x	x	x							
13	x	x	x	x	x						

Figure 4.29 Permitted channel combinations for two different WLANs working in the same AP. (Courtesy of Avaya.)

cochannel interference among different users. The maximum values of power depend on each regulating organism. In the United States, the Federal Communications Commission (FCC) has fixed the limit as 1W; in Europe, the ETSI has established it as 100 mW, equivalent isotropic radiated power (EIRP); in Japan, it is 10 mW/MHz. Due to the lack of a coordinating body governing the use made by various users in the neighboring area within the band, it is essential that they respect the maximum power values established with the aim of not perturbing the communications of other equipment working nearby. In fact, it must be remembered that any RF energy sensed by the radio equipment of the WLAN that is not recognized as potentially generated by some IEEE 802.11b device working in DS spread spectrum mode is considered interference. Particularly, this includes the 802.11b signals generated by equipment working in other channels of the band. The existence of interference leads to the reception of packets with errors, which at the MAC level brings about retransmissions, which in turn brings about a fall in throughput. Given that the equipment also has a fallback mechanism at a lower speed (after two failed transmissions, a NACK is sent), the presence of interference can cause the negotiation of lower bit rates (i.e., passing from 11 Mbps to 5.5, 2, or even 1 Mbps with the consequent reduction in throughput). In general, to avoid this situation, it is necessary to try to maintain a signal level of about 10–12 dB more than the noise level. A menu such as that shown in Figure 4.30 offers the user the ability to monitor at any time the conditions of the link in order to find the optimum site of the equipment.

It ought to be mentioned that we should think about not only the typical sources of interference, such as microwave or the power supply line, but also, for example, multipath propagation. In fact, the use of DS spread spectrum in IEEE 802.11b permits us to combat multipath propagation up to a point, depending on the relationship between delay spread and chip time, by implementing a RAKE structure in the receiver, as most manufacturers do. Note as well that working with spread-spectrum techniques permits us to partially combat the most typical interferences, such as the ovens and electrical motors.

4.5.4 Radio Coverage

Once the noise effects, particularly the interferences and the corresponding limitations, have been analyzed, we are able to briefly describe the aspects to be considered when planning the areas to be covered. For this, it is necessary to know the environment (indoor or outdoor as well as its characteristics),

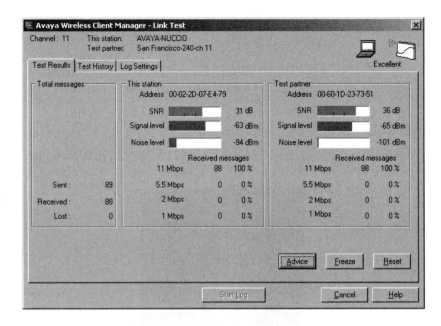

Figure 4.30 Menu that displays the values of signal and noise measured at the local and remote equipment. (Courtesy of Avaya.)

the transmitted power, and sensitivities of the equipment. Table 4.6 shows the values of sensitivity and delay spread [for a *frame error rate* (FER) less than 1%] of the equipment of Avaya at the different working velocities:

Taking these parameters as a reference, Table 4.7 shows some examples of coverage for indoor environments provided by the equipment, assuming a nominal output power of 15 dBm.

The coverage can be extended considerably using a range-extending antenna, such as the one shown in Figure 4.31.

Table 4.6
Receiver Sensitivity and Delay Spread Parameters for WLAN Equipment

Parameters	High Bit Rate (11 Mbps)	Medium Bit Rate (5.5 Mbps)	Standard Bit Rate (2 Mbps)	Low Bit Rate (1 Mbps)
Receiver sensitivity	−83 dBm	−87 dBm	−91 dBm	−94 dBm
Delay spread (FER 1%)	65 ns	225 ns	400 ns	500 ns

Table 4.7
Coverage Areas for Different Indoor Environment Conditions

Environment	High Bit Rate (11 Mbps)	Medium Bit Rate (5.5 Mbps)	Standard Bit Rate (2 Mbps)	Low Bit Rate (1 Mbps)
Open office	160m (525 ft)	270m (885 ft)	400m (1,300 ft)	550m (1,750 ft)
Semi-open office	50m (165 ft)	70m (230 ft)	90m (300 ft)	115m (375 ft)
Closed office	25m (80 ft)	35m (115 ft)	40m (130 ft)	50m (165 ft)

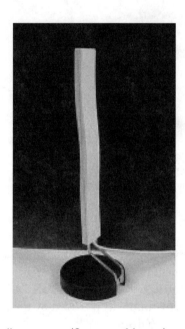

Figure 4.31 Range-extending antenna. (Courtesy of Avaya.)

Connecting a range-extending antenna to the corresponding WLAN card provides a net gain of 2.5 dBi, disabling the antenna card, with an omnidirectional radiation pattern. Finally, it should be mentioned that with the aim of customizing the cell design to the environment needs, the configuration menu of the APs provides the capability of fixing the cell size, modifying the sensitivity thresholds. In this way, it permits, for example, the placement of two APs at a shorter distance, thus responding to the necessities derived from a potentially high concentration of users in determined zones of the areas to be covered.

To finish this practical part relating to the implementation of the IEEE 802.11b infrastructure, we will deal with some aspects related to outdoor environments.

4.5.5   IEEE 802.11 for Outdoor Environment

As was mentioned at the beginning of this chapter, a growing number of cable operators are considering using IEEE 802.11 equipment to provide data services in specific rural areas, where there is a disperse population around an urban nucleus or where the estimated penetration factor is low, making unprofitable the laying of optical fiber or the use of LMDS type systems. Under these conditions, the existence of APs and PCMCIA-type cards, and therefore, low-cost user equipment, along with freedom from operational license applications, make the use of the IEEE 802.11 infrastructure very attractive. To this end, manufacturers such as Avaya offer complete solutions permitting the widespread deployment of wireless IP in outdoor environments.

This deployment makes use of the same equipment as that for indoors, with some additions at the software level. In order to provide better coverage, the manufacturers offer antennas with higher gain than the indoor ones have, especially on the user side. Figure 4.32 shows two typical outdoor antennas, an omnidirectional one and a directional one.

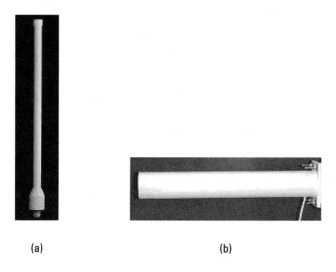

(a)   (b)

Figure 4.32 Outdoor antennas: (a) omnidirectional, and (b) directional. (Courtesy of Avaya.)

The gain for the omnidirectional antenna is 7 dBi, and the gain for the directional one is 14 dBi. The first is associated with the AP, and the second with the user terminal.

To calculate the maximum distances for placing the user terminals from the APs, the following expression for propagation losses can be used:

$$\text{Attenuation (dB) for 2.4 GHz band} = 100 \text{ dB} + 20\log\left(d_{km}\right) \qquad (4.1)$$

Note that an open space propagation model has been assumed. It is also necessary to ensure line of sight between transmitter and receiver along the whole propagation path, particularly for the Fresnel zone, whose radius in the 2.4-GHz band can be obtained from

$$\text{Fresnel Radius (meters)} = 3.4 \cdot \sqrt{d_{km}} + \left(d_{km} / 8.12\right)^2 \qquad (4.2)$$

Table 4.8 shows typical distances separating APs, with omnidirectional antennas, and user equipment for differing regulations and user antenna gains.

Finally, it should be commented that the manufactures provide, through the software configuration of the outdoor equipment, the possibility of limiting the bandwidth assigned to each user, fixing a parameter known as throttle. Typical values of throttle are 64, 128, 256, 384, and 512 Kbps.

Table 4.8
Distances Reachable Outdoors

Bit rate	FCG		ETSI
	14 dBi	24 dBi	14 dBi
11 Mbps	3.5 km	8.5 km	1.2 km
5.5 Mbps	5 km	10 km	1.9 km
2 Mbps	6.5 km	12 km	2.5 km
1 Mbps	8 km	14 km	3.7 km

References

[1] Prasad, N., and A. Prasad, *WLAN Systems and Wireless IP for Next Generation Communications*, Norwood, MA: Artech House, 2002.

[2] Information Technology–Telecommunications and Information Exchange between Systems–Local and Metropolitan Area Networks–Specific Requirements–Part 11: Wireless LAN Medium Access Control (MAC) and Physical Layer (PHY) Specifications, IEEE Std 802.11-1997, June 1997.

[3] Supplement to IEEE Standard for Information Technology–Telecommunications and Information Exchange between Systems–Local and Metropolitan Area Networks–Specific Requirements–Part 11: Wireless LAN Medium Access Control (MAC) and Physical Layer (PHY) Specifications: Higher-Speed Physical Layer Extension in the 2.4-GHz Band, IEEE Std 802.11b-1999, September 1999.

[4] Supplement to IEEE Standard for Information Technology–Telecommunications and Information Exchange between Systems–Local and Metropolitan Area Networks–Specific Requirements–Part 11: Wireless LAN Medium Access Control (MAC) and Physical Layer (PHY) Specifications: High-Speed Physical Layer in the 5-GHz Band, IEEE Std 802.11b-1999, 1999.

[5] Heegard, C., et al., "High-Performance Wireless Ethernet," *IEEE Comm. Mag.*, Vol. 39, No. 11, 2001, pp. 64–73.

[6] Valadas, R., et al., "The Infrared Physical Layer of the IEEE 802.11 Standard for Wireless Local Area Networks," *IEEE Comm. Mag.*, Vol. 36, No. 12, 1998, pp. 107–112.

[7] Rivest, R. L., *The RC4 Encryption Algorithm*, RSA Data Security Inc., 1992 (Proprietary).

[8] Woesner, H., et al., "Power Saving Mechanisms in Emerging Standards for Wireless LAN: The MAC Level Perspective," *IEEE Personal Communications*, Vol. 5, No. 3, 1998, pp. 40–48.

[9] TR 101 683 (V1.1.1), *Broad Radio Access Network (BRAN); High Performance Radio Local Area Networks (HIPERLAN) Type 2; System Overview*, February 2000.

[10] van Nee, R., et al., "New High-Rate Wireless LAN Standards," *IEEE Comm. Mag.* Vol. 37, No. 12, 1999, pp. 82–88.

[11] DTS/BRAN0020004-1 (V0.k), *Broad Radio Access Network (BRAN); High Performance Radio Local Area Networks (HIPERLAN) Type 2 Functional Specification; Data Link Control (DLC) layer; Part 1 – Basic Data Transport Function*, April 2000.

[12] Draft TS 202 761-2 (V0.g), *Broad Radio Access Network (BRAN); High Performance Radio Local Area Networks (HIPERLAN) Type 2 Functional Specification; Data Link Control (DLC) layer; Part 2 – Radio Link Control (RLC) Sublayer*, April 2000.

5

Behavior of the TCP-UDP/IP Protocol Stack over the IEEE 802.11b

5.1 Introduction

As was commented in Chapter 3, the TCP-UDP/IP protocol stack was designed under the hypothesis that the transmission channel supporting these protocols was sufficiently reliable in terms of the probability of receiving a packet with an error at one end of the communication being very low and, therefore, that the loss of a packet was attributable to probable congestion in the network. In this case, the corresponding transport protocols must invoke the flow-control mechanisms that they incorporate.

The wireless channel is a clear example of a transmission medium with a high probability that a transmitted data packet is lost or received with an error. Therefore, it is necessary, when performance-improvement solutions for the TCP-UDP/IP protocol stack are desired in this future wireless scenario denominated 4G, that we first know in detail the behavior of these protocols in their current state over the possible platforms that will make up the heterogeneous network of 4G. In this sense, this chapter provides a description of the way the radio channel influences the performance expected from the different applications running on the corresponding hosts when they communicate using the IEEE 802.11b platform.

The data given in the different sections of this chapter are derived from an exhaustive measurement campaign carried out in an environment such as that described in Figure 5.1. To carry out the campaign, the nonstationary

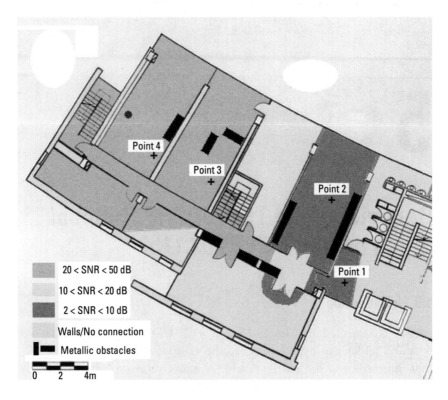

Figure 5.1 Measurement campaign environment.

MT and the fixed AP terminal were used. These were located at various points (from 1 to 4 in Figure 5.1) characterized by some specific SNR intervals.

The wireless cards of both terminals were configured in ad hoc mode, disabling their capacity to negotiate the transmission bit rate depending on the quality of the link. Finally, as measurement and analysis tools, ttcp, tcpdump, tcptrace, xplot, and ethereal programs [1–5] have been used.

5.2   UDP Behavior over IEEE 802.11b

Due to the UDP's characteristics (it follows a connectionless scheme, with no type of error or flow control, constituting a simple connection to join the application layers and IP), it is the ideal choice for characterizing the behavior of the wireless infrastructure based on the IEEE 802.11 protocol. It should be noted that the values obtained with this transport protocol are the maximum values to be expected on this platform and that even with an ideal channel they could never be achieved with TCP.

The analysis of UDP over IEEE 802.11b presented here is structured in three parts. In the first, the performance in what can be considered ideal channel conditions is studied; therefore, the reduction of the binary rate with respect to the nominal value can be attributed to the overhead introduced by the PHY and MAC layers. In the second, the effect of using RTS/CTS frames is evaluated according to what was seen in Chapter 4. Finally, in the third, the behavior of UDP and the wireless platform is evaluated in presence of a channel that introduces errors.

5.2.1 Ideal Channel

In order to assume an error-free channel, the MT is placed in the position identified by point 4 in Figure 5.1, which provides SNRs between 20 and 50 dB. With the aim of having the widest possible characterization, it is convenient to carry out measurements at different bit rates as well as using different frame sizes, which can be fixed using a parameter that the ttcp software has for this purpose (really, what is fixed is the size of the UDP datagram and so indirectly the size of the frame). Table 5.1 shows the average value of the throughput obtained.

With the aim of validating the results that have been obtained, it is convenient to have an analytical support to enable the validation. For this purpose, we will use (5.1), which allows us to obtain the UDP throughput, $Tput_{UDP}$, to be expected in an IEEE 802.11b platform [6]. It defines an effective time, $T_{effective}$, as the time necessary for transmission of the payload

Table 5.1

Measurements of Throughput (Mbps) for a UDP Connection in Ideal Conditions

Payload (Bytes)	R_b (Mbps)			
	11	5.5	2	1
1,472	6.071	3.853	1.691	0.894
996	5.001	3.371	1.574	0.852
740	4.206	2.973	1.466	0.810
484	3.172	2.392	1.285	0.736
228	1.763	1.463	0.917	0.568

(L bytes), without counting the overhead of the headers introduced by all of the layers. Meanwhile, the total time, T_{total}, considers the sum of all of the contributions, both of the headers and of the procedures of the basic access method of DCF.

$$Tput_{UDP}(\text{Mbps}) = \frac{L \cdot 8(\text{bits})}{T_{total}(\mu s)} = \frac{T_{effective}(\mu s)}{T_{total}(\mu s)} \cdot R_b(\text{Mbps}) \qquad (5.1)$$

Figure 5.2 shows the format of the MAC data frame with the size (in bytes) of the headers of all the protocols involved in the communication. In total there are 70 bytes of overhead in each 802.11 frame. Another 192 bits must be added to these 70 bytes corresponding to the preamble and header of the PHY layer, which precede both the data frames and the RTC/CTS or ACK frames seen in Chapter 4. It should be highlighted that these 192 bits are always transmitted, independently of the bit rate at which the rest of the frame is transmitted, at a bit rate of 1 Mbps. For this reason, and with the aim of reducing this overhead, the standard defines an option denominated short format. Figure 5.3 shows both formats and their transmission rates in detail.

Finally, in relation to Figure 5.2, an aspect should be noted that has not been dealt with in previous chapters, the *Subnetwork Access Protocol* (SNAP) field. Given that the majority of the existing LANs are based on Ethernet II technology, with the aim of favoring the compatibility with IEEE recommendations for LANs (802.x family), the SNAP encapsulation is used. This is an extension of the LLC standard of the IEEE 802.2 [7]. The SNAP header, which includes the *service access points'* (SAPs') information, is made up of 3 bytes (see Figure 5.4). The information field is composed of a unique identifier denominated *organizationally unique identifier* (OUI) and the common type used in Ethernet [8].

Note that on adding these 8 SNAP bytes, the Ethernet frame changes its length from 1,500 bytes to 1,492 bytes.

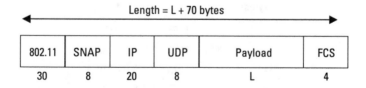

Length = L + 70 bytes					
802.11	SNAP	IP	UDP	Payload	FCS
30	8	20	8	L	4

Figure 5.2 Format of the transmitted frame.

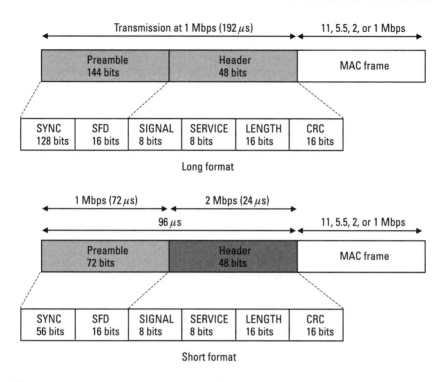

Figure 5.3   Physical preamble format types. (SFD: Start frame delimiter.)

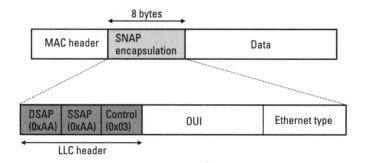

Figure 5.4   SNAP encapsulation.

Continuing with the analysis and following what was also seen in Chapter 4, the DCF method includes diverse procedures that generally bring about a reduction in the throughput, due to the time spent on them (principally associated with the time intervals between frames, backoff intervals, and IEEE 802.11 acknowledgments). Figure 5.5 shows the situation for the case of using a CSMA+ACK procedure.

In Figure 5.5, some specific aspects should be highlighted. The control frames (i.e., the acknowledgments sent by the receiver) are transmitted at the minimum bit rate between the maximum of the set of basic rates (2 Mbps) and the working rate. On the other hand, in a preliminary analysis an average backoff period of 310 μs (15.5 slots $\times$ 20 μs/slot) can be considered. This is because in UDP point-to-point applications, only one station is contending for the medium, so the presence of collisions is negligible. In Chapter 4, however, it was seen that when frames have to be transmitted consecutively, the backoff procedure has to be used, with the aim of not capturing the channel.

Taking into account all of the times appearing in Figure 5.5, (5.1) can be rewritten, obtaining an expression permitting the calculation of the theoretical throughput depending on the working binary rate and the length of the information field.

$$Tput_{UDP}(\text{Mbps}) = \frac{R_b(\text{Mbps})}{1 + \dfrac{70}{L} + \dfrac{R_b(\text{Mbps})}{L} \cdot \left(\dfrac{377}{4} + \dfrac{14}{\min(2, R_b(\text{Mbps}))} \right)} \quad (5.2)$$

Starting from the previous expression, the theoretical throughputs can be calculated for each of the cases in which measurements are made, providing the results shown in Table 5.2.

Note that the deviations between the theoretical values and the measured ones are no greater than 1%, although it should also be noted that the measured throughput, with the wireless platform working at 2 and 1 Mbps,

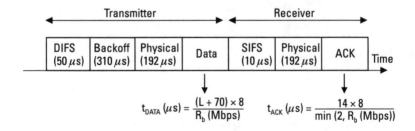

Figure 5.5 Times to be considered in the CSMA+ACK procedure.

Table 5.2
UDP Throughput (Mbps) Obtained Applying (5.2)

Payload (Bytes)	R_b (Mbps)			
	11	5.5	2	1
1,472	6.097	3.857	1.688	0.892
996	5.026	3.375	1.570	0.848
740	4.231	2.978	1.462	0.806
484	3.192	2.396	1.280	0.731
228	1.777	1.467	0.911	0.561

is greater than the theoretical value. The reason for this behavior can be found in the approximation made of the average backoff time. A more detailed analysis of the backoff associated with each frame provides the histogram shown in Figure 5.6. Here, a horizontal window of 620 μs can be seen,

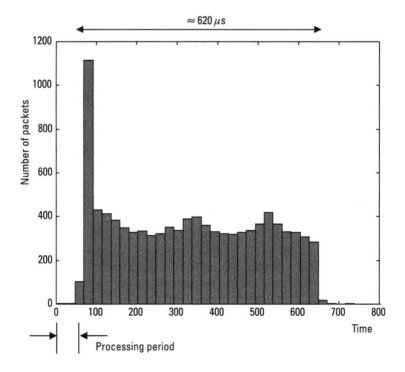

Figure 5.6 Backoff distribution period.

as was indicated previously. Nevertheless, there are two aspects that should be highlighted. The first refers to the fact that the distribution is not strictly uniform, but there are a greater number of frames with a short backoff period. The second has to do with the fact that the smallest backoff value is not zero, as would be expected. In fact there is an offset, of around 30 μs at bit rates of 11 and 5.5 Mbps, due to the processing time (t_{proc}), although this offset reduces at lower bit rates. The consequence of the first of the effects is to reduce the average global backoff and the processing time, as is shown in Tables 5.3–5.6.

Figure 5.7 shows, in graphical form, the effect on the overall performance of each of the procedures used in the DCF scheme.

Table 5.3

Average and Theoretical Throughputs ($R_b = 11$ Mbps)

Payload	Measured Tput$_{UDP}$ (Mbps)	Theoretical (Backoff = 310 μs and t_{proc} = 0) Tput$_{UDP}$	Diff. (%)	Theoretical (Backoff Measured) Backoff + t_{proc}	Tput$_{UDP}$ (Mbps)	Diff. (%)
1,472	6.071	6.097	0.43	319	6.069	0.04
996	5.001	5.026	0.51	319	4.998	0.06
740	4.206	4.231	0.60	318	4.207	0.03
484	3.172	3.192	0.64	318	3.171	0.02
228	1.763	1.777	0.77	318	1.763	0.01

Table 5.4

Average and Theoretical Throughputs ($R_b = 5.5$ Mbps)

Payload	Measured Tput$_{UDP}$ (Mbps)	Theoretical (Backoff = 310 μs and t_{proc} = 0) Tput$_{UDP}$ (Mbps)	Diff. (%)	Theoretical (Backoff Measured) Backoff + t_{proc}	Tput$_{UDP}$ (Mbps)	Diff. (%)
1,472	3.853	3.857	0.11	314	3.852	0.02
996	3.371	3.375	0.13	313	3.371	0.01
740	2.973	2.978	0.15	312	2.975	0.05
484	2.392	2.396	0.18	313	2.392	0.01
228	1.463	1.467	0.27	314	1.462	0.06

Table 5.5
Average and Theoretical Throughputs ($R_b = 2$ Mbps)

Payload	Measured Tput$_{UDP}$ (Mbps)	Theoretical (Backoff = 310 μs and $t_{proc} = 0$) Tput$_{UDP}$ (Mbps)	Diff. (%)	Theoretical (Backoff Measured) Back-off+ t_{proc}	Tput$_{UDP}$ (Mbps)	Diff. (%)
1,472	1.691	1.688	0.20	299	1.690	0.04
996	1.574	1.570	0.23	297	1.574	0.02
740	1.466	1.462	0.29	298	1.466	0.00
484	1.285	1.280	0.42	297	1.285	0.01
228	0.917	0.911	0.64	295	0.918	0.11

Table 5.6
Average and Theoretical Throughputs ($R_b = 1$ Mbps)

Payload	Measured Tput$_{UDP}$ (Mbps)	Theoretical (Backoff = 310 μs and $t_{proc} = 0$) Tput$_{UDP}$ (Mbps)	Diff. (%)	Theoretical (Backoff Measured) Back-off+ t_{proc}	Tput$_{UDP}$ (Mbps)	Diff. (%)
1,472	0.894	0.892	0.23	275	0.894	0.04
996	0.852	0.848	0.45	274	0.851	0.06
740	0.810	0.806	0.51	274	0.810	0.02
484	0.736	0.731	0.70	274	0.736	0.02
228	0.568	0.561	1.19	273	0.568	0.05

Finally, there is a last aspect to analyze. The majority of the applications using UDP as the transport protocol have real-time requirements. Therefore, it is convenient to characterize the average time between consecutive frames and the jitter. In the experimental calculation, the same histograms can be used as were used in the backoff measurements, while (5.3) can be used for the theoretical calculation.

$$\text{Average time between IP packets} = 8 \cdot \frac{\text{Payload (bytes)}}{Tput_{UDP}} \qquad (5.3)$$

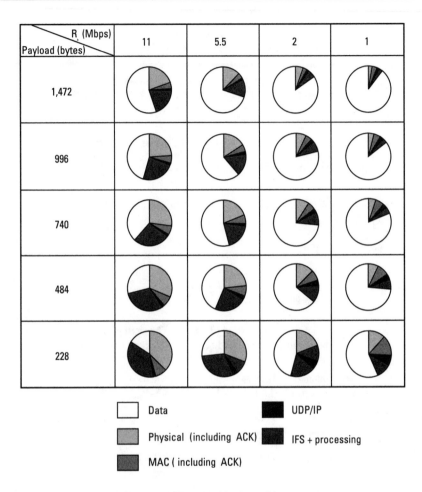

R (Mbps) / Payload (bytes)	11	5.5	2	1
1,472				
996				
740				
484				
228				

Data — □

UDP/IP — ■

Physical (including ACK) — ▨

IFS + processing — ■

MAC (including ACK) — ▨

Figure 5.7 Overall performance in UDP under ideal conditions.

The values obtained with (5.3) correspond exactly to those derived from the measurements. Table 5.7 shows the average values between consecutive UDP packets.

The variance derived from the measurements is shown in Table 5.8, and (5.4) shows how this variance is analytically obtained [9]. Note the great similarity between the two results for the majority of measurements.

$$\text{Delay variance} = \frac{\left(CW_{\max} \cdot \text{slot duration}\right)^2}{12} = \frac{620^2}{12} = 32{,}033\mu s^2 \quad (5.4)$$

Table 5.7
Average Values Between UDP Packets (μs)

Payload (Bytes)	R_b (Mbps)			
	11	5.5	2	1
1,472	1,940	3,057	6,967	13,167
996	1,594	2,363	5,060	9,357
740	1,407	1,990	4,037	7,310
484	1,220	1,619	3,013	5,261
228	1,034	1,247	1,987	3,212

Table 5.8
Variance of Delay Between UDP Packets (μs^2)

Payload (Bytes)	R_b (Mbps)			
	11	5.5	2	1
1,472	34,197	32,827	49,263	55,671
996	32,436	32,715	33,558	36,451
740	32,270	32,420	33,235	34,129
484	32,645	32,573	33,289	34,269
228	32,427	32,179	33,234	34,455

5.2.2 Effect of Access Based on RTS/CTS

Chapter 4 explained the possibility of using an additional mechanism to the basic one, based on the exchange of two short frames (RTS/CTS) that precede the transmission of each data frame. A priori, this method favors those situations in which many terminals are accessing the medium, making the probability of collisions high, or when the hidden terminal phenomenon arises. Due to this, it is obvious that, bearing in mind the development of the measurements in the latter case (with point-to-point communications between two terminals), the use of RTS/CTS is not justified. However, as this is a configurable mechanism when the communications card is enabled, without knowing beforehand how many terminals contend for the medium or whether hidden terminals are present or not, it is useful to know the performance loss due to its use.

In this case, the chronogram of events taking place in the process of transmission/reception varies substantially with respect to what was considered before, as is shown in Figure 5.8.

Tables 5.9–5.12 compare the measured values of throughput with the calculated ones, using both the backoff of 310 μs and that obtained from the measured differences between consecutive frames, following a similar process to that explained in the last section.

Note that the loss of throughput is appreciable when compared with the basic access. Table 5.13 shows the decrease observed in each case analyzed.

From the results shown, it can be deduced that in environments where the presence of hidden terminals is unlikely, and/or the number of stations contending for the medium is not very high, the use of the RTS/CTS

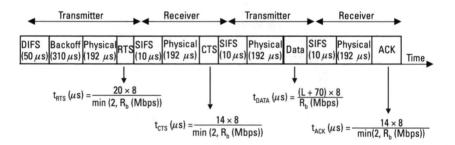

Figure 5.8 RTS/CTS access in DCF.

Table 5.9
Throughput with RTS/CTS (R_b = 11 Mbps)

	Measured	Theoretical (Backoff = 310 μs and t_{proc} = 0)		Theoretical (Backoff Measured)		
Payload	Tput$_{UDP}$ (Mbps)	Tput$_{UDP}$ (Mbps)	Diff. (%)	Backoff + t_{proc}	Tput$_{UDP}$ (Mbps)	Diff. (%)
1,472	4.744	4.765	0.44	321	4.744	0.01
996	3.730	3.749	0.51	321	3.730	0.00
740	3.036	3.053	0.56	320	3.037	0.04
484	2.195	2.209	0.63	320	2.196	0.06
228	1.156	1.164	0.71	320	1.157	0.07

Table 5.10
Throughput with RTS/CTS ($R_b = 5.5$ Mbps)

Payload	Measured Tput$_{UDP}$ (Mbps)	Theoretical (Backoff = 310 μs and $t_{proc} = 0$) Tput$_{UDP}$ (Mbps)	Diff. (%)	Theoretical (Backoff Measured) Backoff + t_{proc}	Tput$_{UDP}$ (Mbps)	Diff. (%)
1,472	3.272	3.278	0.17	315	3.273	0.03
996	2.741	2.747	0.22	315	2.742	0.05
740	2.336	2.342	0.24	316	2.336	0.00
484	1.791	1.796	0.28	317	1.790	0.04
228	1.020	1.023	0.27	315	1.020	0.01

Table 5.11
Throughput with RTS/CTS ($R_b = 2$ Mbps)

Payload	Measured Tput$_{UDP}$ (Mbps)	Theoretical (Backoff = 310 μs and $t_{proc} = 0$) Tput$_{UDP}$ (Mbps)	Diff. (%)	Theoretical (Backoff Measured) Backoff + t_{proc}	Tput$_{UDP}$ (Mbps)	Diff. (%)
1,472	1.568	1.566	0.10	300	1.568	0.03
996	1.422	1.419	0.19	299	1.422	0.01
740	1.293	1.290	0.25	300	1.293	0.03
484	1.089	1.086	0.29	301	1.089	0.04
228	0.721	0.718	0.48	299	0.721	0.05

mechanism leads to a performance loss that means its utilization is not recommendable.

Figure 5.9 shows the different contributions to the overall performance of the IEEE 802.11b protocol, with the RTS/CTS option enabled. In this case, the necessity of implementing the optional short format for the physical header is even more urgent, given that the two frames that are added to the basic scheme mean another two physical headers have to be transmitted, which leads to a notable loss in efficiency. When comparing this figure with Figure 5.7, it can be observed that the differences shown in Table 5.13 are basically due to the increase of the physical overhead.

Table 5.12
Throughput with RTS/CTS ($R_b = 1$ Mbps)

Payload	Measured Tput$_{UDP}$ (Mbps)	Theoretical (Backoff = 310 μs and $t_{proc} = 0$) Tput$_{UDP}$ (Mbps)	Diff. (%)	Theoretical (Backoff Measured) Backoff + t_{proc}	Tput$_{UDP}$ (Mbps)	Diff. (%)
1,472	0.850	0.849	0.17	278	0.850	0.06
996	0.794	0.791	0.34	276	0.794	0.01
740	0.741	0.738	0.41	276	0.741	0.02
484	0.652	0.648	0.59	277	0.652	0.04
228	0.469	0.465	0.94	276	0.469	0.07

Table 5.13
Percentage of Throughput Lost with the RTS/CTS Access Compared to the Basic Access

Payload (Bytes)	R_b (Mbps)			
	11	5.5	2	1
1,472	21.86	15.08	7.27	4.92
996	25.41	18.69	9.66	6.81
740	27.82	21.43	11.80	8.52
484	30.80	25.13	15.25	11.41
228	34.43	30.28	21.37	17.43

5.2.3   Influence of Errors in UDP

This last section dedicated to UDP analyzes the effect that a radio channel, in which the presence of errors is relevant, has on the behavior of applications based on UDP. The analysis focuses on the basic access, given that the quantification of the loss of throughput and the delay due to the errors are the only desired aims.

From the point of view of procedure, the only difference with respect to the two previous sections is that the MT has been placed at the positions marked as point 1 and point 2 of Figure 5.1, in this way guaranteeing the presence of errors (SNR < 10 dB).

At this point, it is helpful to note that the ARQ scheme for detecting errors implemented by the wireless network cards of the measurement

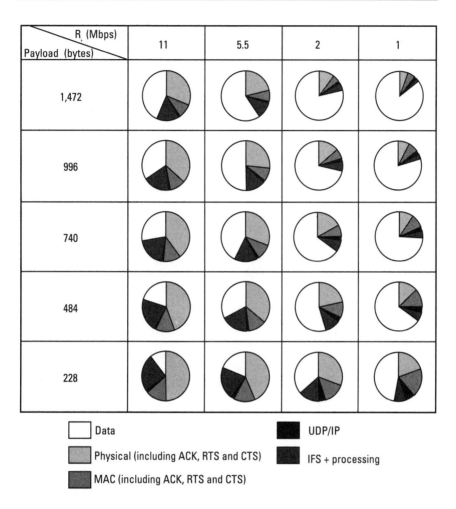

Figure 5.9 Overall performance for UDP in ideal channel conditions with the RTS/CTS mechanism enabled.

platform implies the retransmission of up to a maximum of three times the frame that, having been transmitted, has not been recognized by the end receiver (the standard proposes a default value of seven). This implies that the loss of an IP datagram causes the reception of four consecutive erroneous frames at the 802.11 level, assuming that none have been lost in the radio channel.

In the analysis of applications based on the UDP transport protocol, the parameters of greatest interest for the analysis of the behavior of the wireless platform are the IP loss and the time between consecutive datagrams (given that UDP does not include any error recovery scheme). In order to know the state of the wireless channel at the moment of performing the measurement, the *packet error rate* (PER) is the parameter to be used. The PER is defined as the quotient of the number of erroneous frames that arrive at the wireless card and the total number received. Tables 5.14–5.17 show the data corresponding to a measurement campaign carried out in point 1 (SNR 10 dB), for a transfer of around 10,000 packets from the AP to the MT.

Now, some conclusions relating to the values shown in the previous tables will be given. The PER decreases with the bit rate, for constant packet lengths. This is so because the sensitivity of the wireless card receiver used depends on the working bit rate, according to the data appearing in Table 5.18 [10]. On the other hand, when the bit rate is constant, it can be seen

Table 5.14
UDP Measurements at Point 1 ($R_b = 11$ Mbps)

Payload (Bytes)	Tput$_{UDP}$ (Mbps)	IP Datagrams	MAC Frames	PER (%)	Average Error Burst	Maximum Error Burst	IP Loss (%)
1,472	1.259	6,674	24,168	72.4	6.2	1,736	33.3
996	1.200	7,470	22,937	67.4	4.3	1,644	25.3
740	1.293	8,036	19,346	58.5	4.7	500	19.7
484	1.548	9,273	15,122	38.7	2.7	340	7.3
228	0.999	9,541	13,695	30.3	2.4	448	4.6

Table 5.15
UDP Measurements at Point 1 ($R_b = 5.5$ Mbps)

Payload (Bytes)	Tput$_{UDP}$ (Mbps)	IP Datagrams	MAC Frames	PER (%)	Average Error Burst	Maximum Error Burst	IP Loss (%)
1,472	3.076	9,626	11,436	15.8	5.7	810	3.6
996	2.664	9,611	11,151	13.8	6.6	644	3.1
740	2.673	9,901	10,646	7.0	2.9	252	1.0
484	2.206	9,920	10,441	5.0	3.2	216	0.8
228	1.379	9,962	10,342	3.7	2.3	63	0.4

Table 5.16
UDP Measurements at Point 1 ($R_b = 2$ Mbps)

Payload (Bytes)	Tput$_{UDP}$ (Mbps)	IP Datagrams	MAC Frames	PER (%)	Average Error Burst	Maximum Error Burst	IP Loss (%)
1,472	1.421	9,756	11,327	13.9	3.1	116	2.3
996	1.300	9,666	11,105	13.0	5.0	270	2.7
740	1.386	9,946	10,392	4.3	2.5	38	1.0
484	1.261	9,990	10,061	0.7	2.7	52	0.1
228	0.861	9,936	10,335	3.9	3.6	124	0.6

Table 5.17
UDP Measurements at Point 1 ($R_b = 1$ Mbps)

Payload (Bytes)	Tput$_{UDP}$ (Mbps)	IP Datagrams	MAC Frames	PER (%)	Average Error Burst	Maximum Error Burst	IP Loss (%)
1,472	0.869	9,964	10,104	1.4	2.9	132	0.2
996	0.839	9,995	10,064	0.7	1.5	32	0.0
740	0.796	9,993	10,045	0.5	1.2	12	0.0
484	0.728	10,000	10,051	0.5	1.4	20	0.0
228	0.565	10,002	10,023	0.2	1.9	8	0.0

Table 5.18
WaveLAN Receiver Sensitivity

R_b (Mbps)	Sensitivity (dBm)
11	−83
5.5	−87
2	−91
1	−94

that the PER increases with the length of the packet. The explanation of this behavior is based on the fact that for a constant BER (and therefore a similar

SNR), the PER increases with the packet length. However, it should be noted that this is not always so; it depends on the behavior of other parameters. This is the case of the measurements shown at a bit rate of 2 Mbps, in which on passing from a payload of 484 to 228 octets, it is the average size of the bursts with error (measured in frames) that determines the higher PER obtained for a smaller size of frame. Finally, it should be noted that for a channel with high PER, the throughput reached at 11 Mbps is inferior to that observed working at 5.5 and 2 Mbps. The reason for this is that when the wireless infrastructure experiments a higher PER, the number of retransmissions at 11 Mbps is much greater than in the rest of the cases (observe the MAC frames field of the previous tables, which show the number of MAC frames transmitted). As has been seen in Figure 5.7, the effect of the additional interval times must be added, which is much more noticeable in the performance at higher binary rates.

The great dispersion of the results shown in the previous tables is linked to the extremely changeable behavior of the channel caused by passing different events. In order to obtain, in the best possible way, a more precise characterization, it is convenient to repeat the previous measurements, but in this case fixing the length of the MAC frame at 1,472 bytes (length at which the PER is more appreciable). Tables 5.19 and 5.20 show the results obtained for the five measurements carried out, at points 1 and 2, at the different bit rates and where the number of datagrams sent from the AP to the MT is 15,000.

In both tables, the deviation of the different parameters measured can still be observed for each one of the five measurements. An aspect that has not yet been commented, although it has appeared in previous tables, relates to the average length of error frame burst, which conditions the robustness of the ARQ scheme implemented in the wireless card, which is also defined in the standard. Relating to this, it should be commented that the greater the average burst length, the less likely it is that the retransmission scheme is able to combat these bursts. This is because this scheme envisages a maximum of seven retransmissions (the cards used in the experiments described in this chapter limit this number of retransmissions to three). Over this number, if the receiver has not recognized the frame, the transmitter discards it leading to its loss.

To finish with the study of UDP over IEEE 802.11b, the influence of the errors on the time periods between the correct receptions of two IP datagrams should be characterized. As was mentioned, this aspect is fundamental for those applications that have real-time requirements.

Table 5.19
UDP Measurements (1,472 Bytes of Payload) at Point 1

R_b (Mbps)	Tput$_{UDP}$ (Mbps)	IP Datagrams	MAC Frames	PER (%)	Average Error Burst	Maximum Error Burst	IP Loss (%)
	0.689	6,977	43,630	84	10.22	3,358	54
	0.937	8,301	38,748	79	9.58	1,728	45
	0.944	8,316	38,896	79	9.90	1,292	45
	1.591	11,009	32,311	66	5.36	1,864	27
11	0.955	8,437	38,862	78	8.83	2,384	44
	2.324	14,358	21,916	34	2.10	119	4
	2.198	13,719	21,137	35	3.23	218	9
	2.062	13,625	22,695	40	3.10	262	9
	2.929	14,471	17,922	19	2.84	192	4
5.5	0.473	6,730	39,200	83	12.19	1,576	55
	1.489	14,796	16,536	10	2.46	55	1
	1.460	14,729	16,643	12	2.71	61	2
	1.320	14,404	17,364	17	3.48	141	4
	0.847	12,804	22,323	43	5.46	350	15
2	1.116	13,891	19,602	29	4.09	205	7
	0.429	12,454	21,656	43	4.83	341	17
	0.694	14,309	16,934	15	2.83	45	5
	0.777	14,695	16,332	10	2.70	51	2
	0.810	14,815	15,882	7	2.25	23	1
1	0.749	14,530	16,567	12	3.30	72	4

In the ideal channel case, this time is determined by the backoff mechanism of the 802.11b protocol, and the limits coincide with the maximum and minimum values of the waiting time for a frame (see Figure 5.6). However, taking into account the presence of errors in the wireless channel, it is obvious that this interval will be increased.

Analytically, the average value of the time between consecutive UDP datagrams can be obtained from the quotient of the total time duration of the measurement, t_{tot_meas}, (obtained from the measured throughput,

Table 5.20
UDP Measurements (1,472 Bytes of Payload) at Point 2

R_b (Mbps)	Tput$_{UDP}$ (Mbps)	IP Datagrams	MAC Frames	PER (%)	Average Error Burst	Maximum Error Burst	IP Loss (%)
	3.931	14,063	19,426	28	4.40	625	6
	5.266	14,774	16,416	10	2.88	308	2
	1.509	10,389	31,906	67	9.56	2,295	31
	1.965	12,256	30,039	59	3.78	215	18
11	3.561	14,032	21,040	33	2.99	687	6
	3.212	14,586	16,762	13	4.56	273	3
	1.143	10,810	29,199	63	7.83	3,427	28
	2.051	13,447	22,582	40	4.16	711	10
	3.828	14,989	15,100	1	2.00	32	0
5.5	2.377	13,932	20,677	33	3.74	114	7
	1.603	14,963	15,718	5	1.53	9	0
	1.238	14,441	19,195	25	2.46	168	4
	1.590	14,974	15,838	5	1.36	14	0
	1.355	14,663	17,918	18	2.24	46	2
2	1.641	14,996	15,437	3	1.24	10	0
	0.885	14,995	15,048	0	2.16	12	0
	0.890	14,998	15,078	0	1.66	9	0
	0.888	14,992	15,080	1	2.00	15	0
	0.884	14,999	15,181	1	1.23	5	0
1	0.890	15,002	15,088	0	1.04	2	0

$Tput_{UDP\_meas}$, and from the size of the total information received, $Information_{rx}$) and the number of IP datagrams received, IP_{rx}, according to (5.5).

$$Average\ time\ between\ IP\ packets = \frac{t_{tot\_meas}}{IP_{rx}} = \frac{Information_{rx}}{Tput_{UDP\_meas}} \cdot \frac{1}{IP_{rx}} =$$
$$\frac{IP_{rx} \cdot Payload(bytes) \cdot 8}{Tput_{UDP\_meas}} \cdot \frac{1}{IP_{rx}} = \frac{8 \cdot Payload(bytes)}{Tput_{UDP\_meas}} \quad (5.5)$$

Table 5.21 summarizes the results obtained applying the previous expression and those derived from the measurements carried out.

Finally, Table 5.22 shows the variance associated with the time interval between consecutive IP datagrams and so enables the comparison with the ideal channel case.

5.3 Behavior of TCP over IEEE 802.11

With the aid of what has been seen in previous sections, permitting the characterization of both UDP and the IEEE 802.11b platform under different radio channel conditions, the characterization of the behavior of the TCP

Table 5.21

Average Time (s) Between IP Datagrams

Payload (Bytes)	R_b (Mbps)							
	11		5.5		2		1	
	Measured	Theoretical	Measured	Theoretical	Measured	Theoretical	Measured	Theoretical
1472	9,741	9,353	3,817	3,828	8,274	8,287	13,538	13,551
996	6,644	6,640	2,986	2,991	6,120	6,129	9,498	9,497
740	4,610	4,579	2,203	2,215	4,258	4,271	7,432	7,437
484	2,468	2,501	1,749	1,755	3,059	3,071	5,312	5,319
228	1,832	1,826	1,320	1,323	2,117	2,118	3,228	3,228

Table 5.22

Variance of the Time Between Consecutive UDP Packets (ms^2)

Payload (Bytes)	R_b (Mbps)			
	11	5.5	2	1
1,472	4957.21	1286.33	338.08	367.00
996	2583.19	566.359	554.73	15.52
740	773.69	56.35	12.38	5.03
484	123.58	29.25	5.13	1.97
228	95.94	3.76	16.43	0.31

protocols over the same platform can now be approached. To do this, the same process as with UDP is followed (i.e., the behavior of TCP is analyzed over both the ideal channel and the real channel). Note that in this section, in contrast to what was done before, an analysis of the different procedures of the IEEE 802.11 access protocol will not be contemplated; instead, a detailed study about how TCP reacts to the presence of errors and the implications will be provided.

5.3.1 Ideal Channel

As was explained in Chapter 3, the main difference between the TCP and UDP protocols, in terms of the characterization being done, is that the former implements error- and flow-control mechanisms, based on a cumulative window scheme. In this sense, the station receiving a TCP segment must inform the transmitter of the correct reception by means of acknowledgment segments. The direct consequence of this is that there will no longer be a unique station accessing the medium (as was the case for UDP), but there will be two contending for the channel. This will, in a way, reduce the precision of the data being managed, given that it is no longer possible to establish thoroughly the waiting time of each frame, according to the backoff mechanism implemented by IEEE 802.11b. However, as was also commented previously, the interest of this section is focused on the analysis of the effect of the IEEE 802.11b protocol (and later, the errors in the channel) on TCP.

In the case of the ideal channel, the MT is placed in the position indicated by point 4 (see Figure 5.1), carrying out an FTP transmission of a file of 10 Mbytes, which is captured by the tcpdump tool, for its posterior analysis with tcptrace. The maximum size of the datagram is fixed at 1,500 octets, as was seen in Section 5.2.1; this corresponds to the situation in which the performance is best. Table 5.23 shows the values of the most relevant parameters obtained from the measurements of the four bit rates:

1. The TCP measured throughput, $Tput_{TCP\_meas}$, which is the parameter that characterizes the performance achieved;

2. The number of TCP segments transmitted, $Segments_{TCP}Tx$, which is used to evaluate efficiency loss due to mechanisms used in DCF;

3. The number of acknowledgments sent by the receiver, $ACK_{TCP}Rx$, whose presence reduces the efficiency of the transmitter—while the physical transmission of the TCP level acknowledgments is taking place, the transmitter must be waiting to transmit more data.

Table 5.23

FTP Transfer from the Position Point 4

R_b (Mbps)	Tput$_{TCP\_meas}$ (Mbps)	Segments$_{TCP}$ Tx	ACK$_{TCP}$ Rx
11	5.012	7,245	3,967
5.5	3.269	7,245	4,416
2	1.527	7,245	3,657
1	0.807	7,245	4,354

To understand the reason for the difference in performance achieved by TCP compared with UDP, the influence of the different overheads on the overall performance will now be derived. In this case, as the number of TCP acknowledgments per frame is not the same for all the binary rates (see Table 5.23), the calculations are done using the overall time of the communication, which can be derived from the average throughput and from the quantity of information sent (file size).

Before arriving at the expressions, it is useful to note the problem that occurs when estimating the waiting time associated with the backoff mechanism implemented by the IEEE 802.11b protocol. Bearing in mind that traffic originated by the TCP receiver is noticeably less than that generated by the transmitter, it is assumed that it does not need to invoke the backoff procedure in any of its transmissions. The same value is chosen as was supposed in UDP—that is, 310 μs or 15.5 slots, for the segments that the transmitter generates. This approximation is equivalent to supposing a value of about 10 slots of wait per frame (taking into account the segments with data and the TCP acknowledgments), which is similar to the 8.52 slots given in [11].

Now we will examine the different overheads and their effect on the total throughput. First, the throughput dedicated to the transmission of the physical layer overhead, $Tput_{PHY}$, will be evaluated, including that associated with the transmission of the 802.11 acknowledgments. This is shown in (5.6), where the total time dedicated to the transfer of the file is denoted by t_{tot_trans} and the size of the file by $Size_{file}$ (in bytes).

$$Tput_{PHY} = \frac{(192+192) \cdot Segments_{TCP}Tx}{t_{tot\_trans}} \cdot R_b = 48 \cdot$$

$$\frac{Segments_{TCP}Tx}{Size_{file}} \cdot Tput_{TCP\_meas} \cdot R_b \qquad (5.6)$$

As was seen in the UDP analysis, the waiting mechanisms implemented by the IEEE 802.11b protocol (backoff, *BO*, *DIFS*, and *SIFS*) bring about an appreciable loss in efficiency, see (5.7), where an average waiting time of BO = 310 μs per frame transporting a TCP segment with data is assumed.

$$Tput_{IFS} = \frac{(BO + DIFS + SIFS) \cdot Segments_{TCP}Tx}{t_{tot\_trans}} R_b = \frac{185}{4} \cdot$$

$$\frac{Segments_{TCP}Tx}{Size_{file}} \cdot Tput_{TCP\_meas} \cdot R_b$$

(5.7)

In the same way the throughput loss, $Tput_{MAC}$ due to the overhead of the MAC layer, $Tx_{34bytes}$, is obtained, including the time taken in the transmission of the IEEE 802.11 acknowledgment generated by the receiver station, $Tx_{ACK802.11}$, and in the transmission of the SNAP encapsulation, Tx_{SNAP}.

$$Tput_{MAC} = \frac{\left(Tx_{34\,bytes} + Tx_{ACK\,802.11} + Tx_{SNAP}\right) \cdot Segments_{TCP}Tx}{t_{tot\_trans}} \cdot R_b =$$

$$2 \cdot \frac{Segments_{TCP}Tx}{Size_{file}} \cdot \left(21 + 7 \cdot \frac{R_b}{\min(R_b,2)}\right) \cdot Tput_{TCP\_meas}$$

(5.8)

At layer 4, the header added by the transport layer is considerably greater than the eight octets introduced by UDP. The basic format of the TCP header is 20 bytes, to which 12 more must be added for transmitting the information corresponding to the *timestamp* option, which is enabled in all the measurements presented here. Thus, the total overhead is 52 octets (32 of TCP and 20 of IP) and (5.9) shows the throughput loss, $Tput_{OHEAD}$, due to the overhead of this layer 4, $Tx_{52bytes}$.

$$Tput_{OHEAD} = \frac{Tx_{52\,bytes} \cdot Segments_{TCP}Tx}{t_{tot\_trans}} \cdot R_b =$$

$$52 \cdot \frac{Segments_{TCP}Tx}{Size_{file}} \cdot Tput_{TCP\_meas}$$

(5.9)

The only thing that now remains to evaluate the performance loss of the transmitter due to the time it is detained during the period when the TCP receiver is transmitting the layer 4 acknowledgments necessary in the cumulative sliding window scheme. From the point of view of IEEE 802.11b, the transmission of these segments generates an identical transmission mechanism to the transmission of a data segment by the transmitter, including the MAC layer ACK. The only difference is that, as was previously seen, it is supposed that the transmission of TCP acknowledgments does not need to wait due to the backoff mechanism. Equation (5.10) permits the evaluation of the part of the total throughput that is destined to the transmission of the TCP ACKs, $Tput_{TCP\_ACK}$.

$$Tput_{TCP\_ACK} = \left(94 + R_b \cdot \left(\frac{111}{2} + \frac{14}{\min(2, R_b)} \right) \right) \cdot \frac{ACK_{TCP}Rx}{Size_{file}} \quad (5.10)$$
$$\cdot \, Tput_{TCP\_meas}$$

The sum of all of the previous contributions must be the same as the working bit rate in each case. If the calculations are done for each case, the reader will observe that the difference between the measured and the analytically derived values is less than 2%. This difference mainly originates from the approximation that was made about the waiting time per frame.

Finally, Figure 5.10 shows the individual contribution of the overheads to the overall performance, identified and quantified previously, for the four working bit rates.

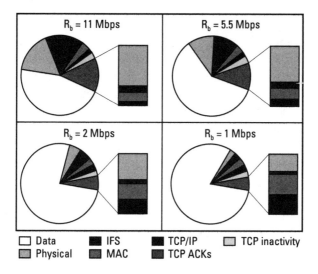

Figure 5.10 Overall performance of TCP under ideal conditions.

5.3.2   Influence of Errors on TCP

The behavior of TCP over the IEEE 802.11b platform in the presence of a radio channel exhibiting a high BER is radically different to that observed in UDP. So, for example, the IP loss will now be zero in any case due to the persistent error-control scheme incorporated in TCP, independent of the fact that at MAC level the same frame is only retransmitted a maximum of four times. As a consequence of the error-control scheme's persistence, the parameter that will be penalized most is the average throughput at TCP level. Tables 5.24 and 5.25 reflect this, showing some of the results obtained for the measurements carried out in positions where the presence of errors is assured, namely, Points 1 and 2 in Figure 5.1. Note that, in specific cases and despite working in an unfavorable position, the TCP measured throughputs reached are similar to those obtained in the ideal channel case. As seen in the UDP analysis, this is further conformation of the extreme variability of the channel, due to the fact that this type of infrastructure, unlike our case, is not usually used in positions ensuring line of sight at all times.

In these conditions, there are three main reasons contributing to the degradation of the TCP throughput:

- The retransmissions generated to recover lost segments;

- The presence of erroneous frames (nonzero PER), which means that the number of MAC frames transmitted is greater than the number of TCP segments;

- The appearance of a period of inactivity of the TCP transmitter entity. This period is even shown in an ideal channel, where it was ignored due to its little relevance in the interpretation of the results. However, for the channels of interest here, its relevance may even be a dominant factor in the possible degradation of the average throughput. The proportion of throughput spent in these inactive periods, T_{IDLE}, is reflected in the previous tables through the parameter $Tput_{IDLE}$.

In relation to this last aspect, in the next section the complete development justifying the appearance of the inactive periods is presented. This development is presented for two reasons: to be able to understand how TCP (and thus its throughput) is affected by the presence of a radio channel exhibiting a high BER, and to enable the reader to understand the error and flow

Table 5.24

TCP Throughputs (Mbps) Measured at Point 1

R_b	Tput$_{TCP\_}$ meas	Tput$_{OHEAD}$	[1]Tput$_{802.11}$	[2]Tput BACKOFF	[3]Tput$_{RTX}$	Tput$_{TCP\_}$ ACK	Tput$_{IDLE}$	T$_{IDLE}$ (s)
	2.906	0.122	1.68	0.99	0.84	0.85	3.616	9.76
	0.707	0.030	0.42	0.25	0.25	0.24	9.109	101.05
	4.488	0.186	2.38	1.40	0.51	1.59	0.436	0.76
	0.501	0.021	0.28	0.16	0.10	0.22	9.719	152.15
11	4.586	0.190	2.30	1.36	0.03	1.58	0.959	1.64
	3.280	0.136	0.86	0.48	0.01	0.57	0.162	0.78
	2.850	0.118	0.77	0.44	0.16	0.51	0.646	3.56
	0.686	0.029	0.19	0.11	0.06	0.13	4.298	98.29
	1.074	0.045	0.30	0.17	0.09	0.33	3.502	51.15
5.5	2.655	0.110	0.72	0.41	0.15	0.59	0.860	5.08
	0.758	0.032	0.09	0.04	0.04	0.07	0.977	55.58
	0.768	0.032	0.09	0.04	0.03	0.07	0.969	54.44
	1.302	0.055	0.15	0.07	0.03	0.11	0.278	9.22
	1.340	0.056	0.15	0.07	0.01	0.11	0.260	8.36
2	0.389	0.016	0.04	0.02	0.02	0.04	1.468	162.80
	0.387	0.017	0.03	0.01	0.02	0.03	0.505	112.60
	0.786	0.033	0.06	0.02	0.00	0.05	0.045	4.95
	0.352	0.015	0.03	0.01	0.01	0.03	0.558	136.71
	0.792	0.033	0.06	0.02	0.01	0.05	0.034	3.67
1	0.771	0.032	0.06	0.02	0.01	0.05	0.053	5.89

1. Tput$_{802.11}$ represents the throughput loss due to all IEEE 802.11b mechanisms except the backoff period.

2. Tput$_{BACKOFF}$ represents the throughput loss due to the backoff mechanism.

3. Tput$_{RTX}$ represents the throughput loss due to the TCP retransmissions.

control mechanisms explained in Chapter 3, which are manifested here to their highest degree.

5.3.2.1 Analysis of the Idle Times

Before approaching the detailed study of how TCP undergoes idle times with such high values as those shown in Tables 5.24 and 5.25, it is necessary to provide some basic knowledge about how the operating system used, Linux [12], manages the retransmission timeouts.

Table 5.25

TCP Throughputs (Mbps) Measured at Point 2

R_b	$Tput_{TCP\_meas}$	$Tput_{OHEAD}$	$^1Tput_{802.11}$	$^2Tput_{BACKOFF}$	$^3Tput_{RTX}$	$Tput_{TCP\_ACK}$	$Tput_{IDLE}$	$T_{IDLE}(s)$
	4.67	0.194	2.33	1.38	0.000	1.27	1.158	1.95
	1.98	0.083	1.15	0.68	0.034	0.67	6.081	24.09
	4.84	0.201	2.43	1.43	0.002	1.36	0.714	1.16
	4.47	0.185	2.25	1.33	0.002	1.50	1.215	2.13
11	4.89	0.203	2.46	1.45	0.000	1.49	0.467	0.75
	3.31	0.137	0.24	0.49	0.000	0.00	1.328	6.29
	2.38	0.099	0.18	0.37	0.015	0.79	1.577	10.39
	1.86	0.077	0.14	0.28	0.003	0.82	2.266	19.11
	3.27	0.135	0.23	0.48	0.000	1.10	0.271	1.30
5.5	3.18	0.132	0.23	0.47	0.005	1.09	0.359	1.77
	1.49	0.062	0.11	0.08	0.015	0.00	0.242	7.00
	1.51	0.063	0.11	0.08	0.000	0.18	0.062	1.77
	1.48	0.062	0.11	0.08	0.010	0.18	0.087	2.52
	1.51	0.063	0.11	0.08	0.005	0.18	0.056	1.61
2	1.53	0.063	0.11	0.08	0.000	0.18	0.039	1.11
	0.805	0.033	0.06	0.02	0.000	0.00	0.082	8.84
	0.806	0.033	0.06	0.02	0.000	0.05	0.027	2.90
	0.807	0.033	0.06	0.02	0.000	0.05	0.025	2.71
	0.801	0.033	0.06	0.02	0.000	0.05	0.032	3.47
1	0.807	0.033	0.06	0.02	0.000	0.06	0.025	2.63

1. $Tput_{802.11}$ represents the throughput loss due to all IEEE 802.11b mechanisms except the backoff period.

2. $Tput_{BACKOFF}$ represents the throughput loss due to the backoff mechanism.

3. $Tput_{RTX}$ represents the throughput loss due to the TCP retransmissions.

The Linux kernel manages all of the TCP connections established using a data structure (*struct tcp_opt*), containing all of the variables needed for the management of the procedures used in them. The most important are described in the following list [13]:

- *packets_out.* These are the TCP segments that have been sent and have not yet been acknowledged by the receiver.

- *fackets_out.* These are TCP segments that have been acknowledged with SACK information.

- *retrans_out.* These are retransmissions generated by the fast retransmit mechanism.

- *retransmits.* This is the number of retransmissions triggered by expiration of the RTO.

- *srtt.* This is smoothed RTT (scaled by eight, to facilitate the implementation).

- *mdev.* This is the RTT standard deviation (scaled by four).

- *rto.* This is the current value of timeout to be applied in the retransmissions.

- *backoff.* This is used when a segment must be retransmitted more than once.

- *snd_cwnd.* This is the current value of the congestion window (expressed in segments).

- *snd_ssthresh.* This is threshold managed in the slow start procedure.

- *snd_cwnd_cnt.* This is the variable used in the congestion-avoidance algorithm to implement the linear increment (when the slow start zone limit has been surpassed).

As mentioned before, the overall measurements presented for the TCP protocol were with the TCP timestamp option enabled. This conditions the TCP timeout retransmission mechanism to a great extent, as the estimation of the values for the RTT and its standard deviation (used later in the generation of the RTO) follows a very different process than the usual one (see Chapter 3). When the timestamp option is selected, the transmitter places (in the options field of the TCP header of each segment sent) the current temporal value, and the receiver reflects it in the corresponding acknowledgments. Starting from the reception of an ACK acknowledging new data, the TCP transmitter entity only has to evaluate a simple subtraction to obtain a suitable measurement of RTT, which will be used in the algorithms that calculate the smoothed values of its average value and standard deviation. As well as improving the RTT estimation mechanism, the incorporation of this procedure simplifies the implementation of the TCP transmitter entity.

In Linux, the function that manages the updating of the RTT values is denominated *tcp_ack_saw_tstamp*, and it is called every time that a segment with the ACK indicator activated is received. Figure 5.11 shows the flow diagram of this function.

As can be seen in Figure 5.11, only those acknowledgments that acknowledge new data are used for updating the RTT values. On the other hand, and this is an extremely important point in the analysis carried out later, when the variable retransmitted is not zero (a packet has been retransmitted for timeout), it will not be zero again until the number of packets that were in the network at that moment (*packets_out*) is zero (i.e., until the receiver has confirmed all the pending data). This fact is fundamental because, following Nagle's algorithm [14], a TCP entity cannot transmit new packets while the *retransmits* value is nonzero (it is retransmitting).

In the same way, Figure 5.11 includes three procedures that do not appear in detail: the calculation of RTT, RTO, and the limit of this last parameter. These operations are carried out by three functions in the TCP

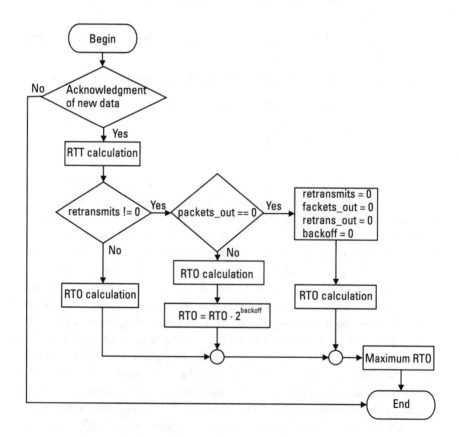

Figure 5.11 Flow chart of the function *tcp_ack_saw_tstamp*.

implementation, which are followed in the kernel of Linux and which are, respectively, *tcp_rtt_estimator, tcp_set_rto,* and *tcp_bound_rto.* The flow diagrams of the three functions are shown in Figure 5.12. The first, which estimates the average value and the variance of the RTT, is called from *tcp_ack_saw_tstamp,* which provides as a parameter the difference between the current time of the machine and the echo that the receiver sends it in the corresponding field in the TCP header options. It should be remembered that, to favor the implementation, the values managed for *srtt* and *mdev* are scaled by eight and four, respectively [15]. It is easy, however, to notice that the value used for the calculation of the RTO is adequate—given by (3.3).

Despite recommending the use of (3.3) for the calculation of the time to be applied in the retransmission timeout, it can be observed that in the TCP implementation in the Linux kernel, the result of it is multiplied by a factor depending on the congestion window of the connection at any moment. The *tcp_bound_rto* function limits the timeout to be applied, both lower (the minimum timeout that can be used is 200 milliseconds) and

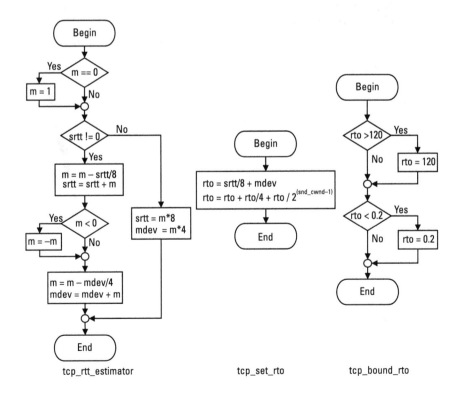

Figure 5.12 Algorithms for the calculation of the RTT and RTO in TCP.

higher (it cannot wait more than 2 minutes for retransmission). Finally, each time a segment is transmitted, a timer is initialized, with the current value of the RTO. If this timer expires on any occasion, the process of retransmission of that packet is triggered.

Now we are in a position to begin the analysis of the idle times of any capture. For this a measurement that shows an idle time of 57 seconds is chosen. The trace of this measurement is shown in Figure 5.13. Note that this interruption in the transmission of segments, during a period of 57 seconds, is the predominant cause of the degradation of the TCP throughput in this measurement at 11 Mbps.

In the same way, Figure 5.14 shows the interchange of segments previous to the interruption of the transmission, which is the direct cause of it.

In order to facilitate the analysis, in the previous diagram a TCP segment sequence number that starts with a value of 1 was used. The notation used in the acknowledgments is the same as that used in the usual function of the protocol; when the receiver sends an *ack8*, it indicates that the last segment received correctly was the seventh, and the next to be expected is the eighth. The time reference is not exactly as is shown, as the effect of the IEEE 802.11b protocol must be taken into account. This provokes the interchange

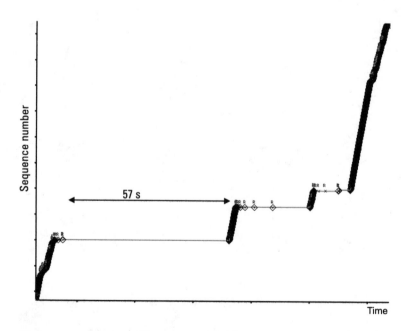

Figure 5.13 Interruption of 57 seconds in a TCP connection.

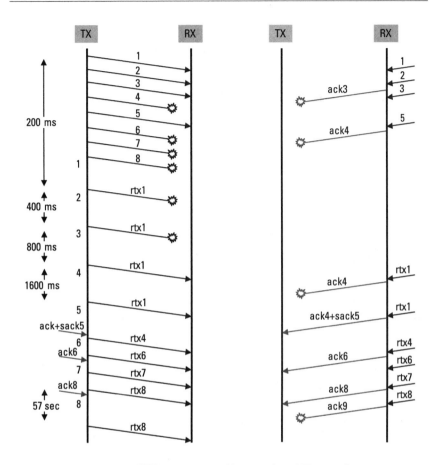

Figure 5.14 Interchange of TCP segments and interruption of 57 seconds.

of frames so they are not in the strict order in which they are generated by the two corresponding TCP entities. On the other hand, the lost segments are really a consequence of the existence of four consecutive erroneous MAC frames, as has been mentioned repeatedly.

Table 5.26 shows the evolution of the variables that must be considered during the interchange of the segments shown in Figure 5.14. The values of *srtt* and *mdev* can be obtained from the ethereal application and a simple program developed for this purpose. The rest of the parameters can be derived from the analysis of the capture. Thus, for example, it can be observed that the transmitter transmits eight packets at the start and then waits to receive confirmation from the receiver. In this way, the value of *snd_cwnd* at the start

Table 5.26
Evolution of the TCP Connection Until the 57-Second Interruption

Step	Event	srtt (ms)	mdev (ms)	snd_ cwnd	snd_ ssthresh	Retrans- mits	Backoff	packets_ out
1	Start	30	10	8	?	0	0	8
2	Rtx 1 (tmo)	30	10	1	4	1	1	8
3	Rtx 1 (tmo)	30	10	1	4	2	2	8
4	Rtx 1 (tmo)	30	10	1	4	3	3	8
5	Rtx 1 (tmo)	30	10	1	4	4	4	8
6	Rx ack(3)	400	750	2	4	4	4	5
7	Rx ack(6)	350	660	3	4	4	4	3
8	Rx ack(8)	310	580	4	4	4	4	1

is of eight segments. From this moment, the variation of the rest of the variables is simple. In the columns corresponding to the values of the RTT, the values have been used without scaling and the step column corresponds to the numbers shown in Figure 5.14.

After the retransmission of segment one, due to expiration of the timeout, the slow start procedure is invoked, the value of *snd_cw*nd passes to 1, and the threshold of slow start, *snd_ssthresh*, passes to half of the previous congestion window, *snd_cwnd*, (that is, four). The successive retransmissions of the same segment cause the values of the variables *backoff* and *retransmits* to increase gradually. In step six, the acknowledgment of segments 1 and 2 is received, *ack3*, as well as that of 5 (through the SACK option). At this moment, the transmitter retransmits two packets from its retransmission queue, which are triggered by a function of the TCP implementation (*tcp_xmit_retransmit_queue*) that is called each time new data is recognized when in conditions to retransmit. No special mention has been made of this, as it does not affect the variables whose evolution is being analyzed. The arrival of the three acknowledgments (*ack3*, *ack6*, and *ack8*) does, however, cause the updating of the variables that manage the RTT, as is shown in Table 5.26. Furthermore, it reduces the number of packets that are transmitted without confirmation. In contrast to what might appear logical, the variables *retransmits* and *backoff* do not vary, even though the segment that caused their increase has now been recognized, as was mentioned previously when describing the algorithm of the function that manages the timestamp procedure (see Figure 5.11). In other words, until *packets_out* is zero, neither *retransmits* nor *backoff* will be updated.

It remains to be proven that the value of the RTO applied in the retransmission of segment 8, obtained from the data in Table 5.26, corresponds to what was measured (approximately 57 seconds). If the flow chart in Figure 5.11 is analyzed once more, the procedure followed on receiving *ack8* can be deduced. First, as new data are confirmed, the values of RTT are estimated (shown in Table 5.26). Next, as *retransmits* and *packets_out* have a nonzero value, the RTO is calculated, calling the function *tcp_set_rto*, to later apply the corresponding backoff. Proceeding in this way, the result for the RTO is shown in (5.11).

$$RTO = \left(srtt + 4 \cdot mdev\right) \cdot \left(1 + 0.25 + \frac{1}{2^{snd\_cwnd-1}}\right) \cdot 2^{backoff} =$$

$$\left(310 + 4 \cdot 580\right) \cdot \left(1 + 0.25 + \frac{1}{2^3}\right) \cdot 2^4 = 57,860 \text{ ms}$$

(5.11)

As can be seen, the value obtained coincides with that which was measured in the captures.

Finally, to finish this section and the chapter, the analysis of a second case is included corresponding to the fourth measurement done at 11 Mbps (Table 5.24), in which an interruption of 120 seconds once again severely degrades the TCP throughput. Given that this case is very similar to the previous one, less detail is given in the explanation. Nevertheless, its analysis is interesting, given that the time of inactivity reached is the maximum permitted by the TCP implementation being used (see the *tcp_bound_rto* function in Figure 5.12). Figure 5.15 illustrates the great influence on TCP throughput of the period of inactivity of 120 seconds.

The interchange of TCP segments caused by the previous situation is shown in Figure 5.16.

The evolution of the corresponding variables during the previous interchange of segments is shown in Table 5.27.

In this case, the value of the RTO derived is over 120 seconds, as is shown in (5.12), so the function limiting this value fixes it at the 120 seconds observed.

$$RTO = \left(srtt + 4 \cdot mdev\right) \cdot \left(1 + 0.25 + \frac{1}{2^{snd\_cwnd-1}}\right) \cdot 2^{backoff}$$

$$= \left(700 + 4 \cdot 1,360\right) \cdot \left(1 + 0.25 + \frac{1}{2^2}\right) \cdot 2^5 = 322,350 \text{ sec}$$

(5.12)

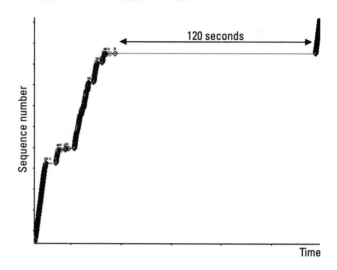

Figure 5.15 Interruption of 120 seconds in a TCP connection.

Although we could continue with numerous analyses that would complement the results presented here, we think that at this point the reader will have a clear idea of some of the limitations of the wireless platforms and the transport protocols in practice. Among these, maybe the most preoccupying is that shown by TCP, especially when the wireless channel displays certain BERs. We believe this behavior is one of the limitations with greatest implications in the path toward a true deployment of 4G, whereby the only agglutinating element of the different infrastructures, WPAN, WLAN, personal, or cellular networks, is precisely the TCP-UDP/IP protocol stack.

5.4 Conclusions

Given the large quantity of information provided in this chapter, some conclusions will be highlighted. First, when planning to use IEEE 802.11b for transporting UDP streams, it has been shown that due to the absence of forward error correction mechanisms defined in this standard, the loss of datagrams in adverse propagation, which implies the jitter value increase, can seriously degrade the information received at the other end of the communication. In the same way, when the transport done is based on TCP protocol, the continual loss of TCP segments provokes the appearance of prolonged periods of inactivity leading to drastic falls in TCP throughput.

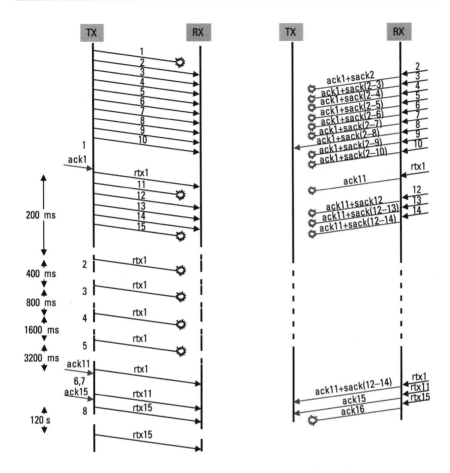

Figure 5.16 Interchange of segments and interruption of 120 seconds in TCP connection.

It is important to remark once more that the main reason is because the TCP-UDP/IP stack was designed under the hypothesis that the communications infrastructure, supporting them exhibits a very low packet loss rate, which justifies the results already shown. As it is well known, wireless platforms are far from this assumption. It is in this context that a solution that allows us to cope with this behavior must be identified. In Chapter 7, the design of an intermediate layer between the wireless infrastructure and the IP layer is proposed. Furthermore, this scheme will be proposed as a candidate towards the 4G paradigm.

Table 5.27
Evolution of the TCP Connection Until the 120-Second Interruption

Step	Event	srtt (ms)	mdev (ms)	snd_ cwnd	snd_ ssthresh	Retransmits	Back- off	packets_out
1	Start	30	10	10	?	0	0	10
2	Rtx 1 (tmo)	30	10	1	5	1	1	10
3	Rtx 1 (tmo)	30	10	1	5	2	2	10
4	Rtx 1 (tmo)	30	10	1	5	3	3	10
5	Rtx 1 (tmo)	30	10	1	5	4	4	10
6	Rtx 1 (tmo)	30	10	1	5	5	5	10
7	Rx ack(11)	800	1,550	2	5	5	5	5
8	Rx ack(15)	700	1,360	3	5	5	5	1

References

[1]   ttcp: http://renoir.csc.ncsu.edu/ttcp/.

[2]   tcpdump: http://www.tcpdump.org/.

[3]   tcptrace: http://www.tcptrace.org/.

[4]   xplot: http://www.tcptrace.org/useful.html.

[5]   ethereal: http://www.ethereal.com/.

[6]   Bing, B., "Measured Performance of the IEEE 802.11 Wireless LAN," *Proc. 24th Conference on Local Computer Networks*, Lowell, MA, October 17–20, 1999, pp. 34–42.

[7]   Postel, J., and J. Reynolds, "A Standard for the Transmission of IP Datagrams over IEEE 802 Networks," RFC 1042, February 1998.

[8]   Ferrero, A., *The Eternal Ethernet*, 2nd ed., Reading, MA: Addison-Wesley, 1999.

[9]   Papoulis, A., *Probability, Random Variables, and Stochastic Processes*, 3rd ed., New York: McGraw-Hill, 1991.

[10] Lucent Technologies, "WaveLAN/IEEE Turbo 11 Mb PC Card Quick Installation Guide," September 1999.

[11] Kamerman, A., and G. Aben, "Net Throughput with IEEE 802.11 Wireless LANs," *Wireless Communications and Networking Conference,* Chicago, IL, September 23–28, 2000, pp. 747–752.

[12] Stevens, W. R., *TCP/IP Illustrated: The Implementation,* Reading, MA: Addison-Wesley, 1995.

[13] Gleditsch, A. G., and P. K. Gjermshus, "Linux Cross-Reference," http://lxr.linux.no/.

[14] Nagle, J., "Congestion Control in IP/TCP Internetworks," RFC 896, January 1984.

[15] Jacobson, V., and M. Karels, "Congestion Avoidance and Control," *Computer Communication Review,* Vol. 18, No. 4, August 1988, pp. 314–329.

6

WPANs

6.1 Introduction

A PAN is a network solution that enhances our personal environment, either work or private, by networking a variety of personal and wearable devices within the space surrounding a person, and providing the communication capabilities within that space and with the outside world.

PAN represents the person-centered network concept (see Figure 6.1), which will allow the person to communicate with his or her personal devices close to him or her (e.g., personal digital assistants, webpads, organizers, handheld computers, cameras, and head-mounted displays) and to establish the wireless connections with the outside world [1–5].

Wireless communications experienced dramatic growth within the last decade (GSM, IS-95, GPRS and EDGE, UMTS, and IMT-2000). The siege for higher data bit rates resulted in new wireless systems and networks solutions. The advantages of the wireless world and the desire for higher mobility caused the replacement of fixed connections to the communication networks and the development of different PAN solutions, which will change the concept of a terminal to a person and his or her personal space. PAN is a new member of the GIMCV family.

It will cover the personal space surrounding the person within the distance to which the voice reaches. It will have a capacity in the range of 10 bps to 10 Mbps (see Figure 6.2). Existing solutions (i.e., Bluetooth) operate in the unlicensed *instrumental, scientific, and medical* (ISM) frequency band of

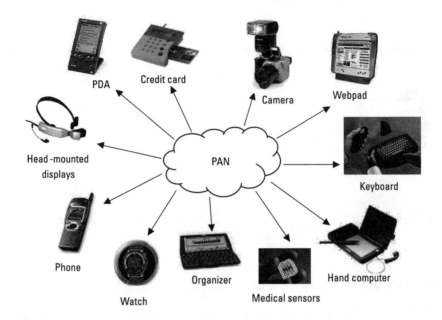

Figure 6.1 PAN as a network solution.

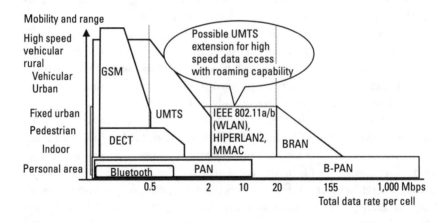

Figure 6.2 Where PAN belongs.

2.4 GHz (see Figure 6.3). Future PAN systems will operate in the unlicensed bands of 5 GHz and perhaps higher. PAN is a dynamic network concept, which will demand appropriate technical solutions for the architecture, protocols, management, and security.

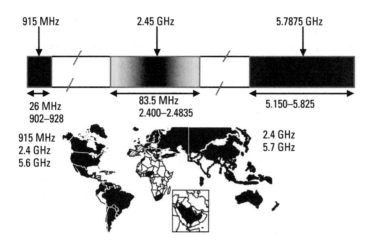

Figure 6.3 Unlicensed frequency bands.

Existing concepts of PAN are mentioned in Section 6.2. Section 6.3 presents an overview of Bluetooth. Different aspects of PAN have been introduced in Sections 6.4–6.13. Section 6.14 compares WLANs and WPANs. Finally, conclusions are given in Section 6.15.

6.2 Existing Concepts

Person-centered network concepts were developed from a Massachusetts Institute of Technology (MIT) idea [6] from 1995 to use the intrabody electrical currents for communication between the devices attached to the human body. It was first accepted by IBM Research [7] and afterwards experienced many variations developed by different research institutions or companies. The different variety of PAN solutions include:

- Oxygen project (MIT) [8];
- Pico-radio [9];
- Infared Data Association (IrDA) [10];
- Bluetooth [11];
- IEEE 802.15 [12].

The Bluetooth concept, originally developed as a cable replacement, is becoming widely accepted, and some of its ideas are incorporated in the ongoing IEEE 802.15 standards concerning PANs.

6.3   Overview of Bluetooth

This section describes the Bluetooth standard, with particular attention to its baseband, LMP, and L2CAP specifications.

Bluetooth is defined to support wireless communications; it was developed by the Bluetooth Special Interest Group (SIG), constituted mainly of:

- Nokia Mobile Phones;
- Ericsson Mobile Communications AB;
- IBM Corporation;
- Intel Corporation;
- Toshiba Corporation.

The Bluetooth specifications were published in 1999; nowadays, some vendors commercialize products that implement this system.

6.3.1   Bluetooth General Architecture

Bluetooth is designed to be used in a short-range radio link between two or more mobile stations. The system provides a point-to-point connection between two stations or point-to-multipoint connection where the medium is shared by several stations. We then have a *piconet,* where two or more units share the same medium.

In a piconet, one station acts as master, and the others as slaves. In effect, the names *master* and *slave* refer to the protocol used on the channel: Any Bluetooth unit (all units are identical) can assume one of the two roles when required. The master is defined as the unit that initiates the connection (toward one or more slave units). A piconet can have one master, and up to seven slaves can be in an active state. *Active state* means that a unit is communicating with a master; the station can stay in a *parked state* if it is synchronized to the master, but it is not active on the channel. Both active and parked station are controlled by the master.

A slave unit can be synchronized with another piconet: A station that is master in one piconet can be slave in another one. In this way, multiple piconets with overlapping coverage, which are not time or frequency synchronized, constitute a *scatternet*. These different scenarios are synthesized in Figure 6.4.

The main characteristics of the Bluetooth system are presented in Table 6.1. The Bluetooth system employs a time-slotted access method. A packet can use up to five slots but must have at least one slot. Bluetooth system may transport an asynchronous data channel, up to three simultaneous synchronous voice channels, or a channel that simultaneously supports asynchronous data and synchronous voice.

The different types of links supported by Bluetooth are:

- A 64-Kbps synchronous link in each direction for voice channel;
- Maximal 723.2 Kbps asymmetric in one direction (still up to 57.6 Kbps in the return direction) or 433.9 Kbps symmetric for asynchronous link.

6.3.2 Bluetooth Protocol Reference Model

Figure 6.5 shows a correspondence between the Bluetooth's protocol stack and the standard OSI stack.

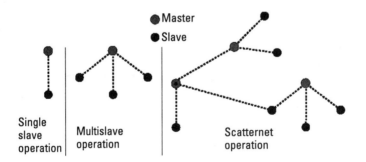

Figure 6.4 Different interconnection between stations.

Table 6.1
Bluetooth System Characteristics

Spectrum	2.4 GHz
Maximum physical rate (symbol rate)	1 Mbps
Access method	TDMA/FDMA-TDD

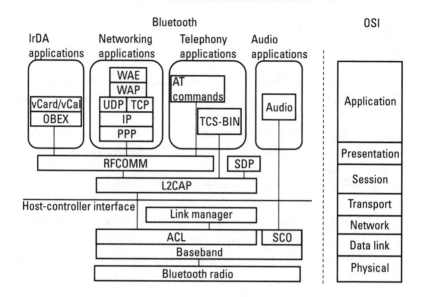

Figure 6.5 Bluetooth protocol reference model.

From Figure 6.5, we observe the existence of both Bluetooth-specific protocols, like LMP and L2CAP, and non-Bluetooth-specific protocols like PPP, IP, and TCS BIN. In Bluetooth, we can distinguish four groups of protocols according to their purpose:

1. *Bluetooth core protocols (baseband, LMP, L2CAP, and SDP):* This group consists of specific Bluetooth protocols developed by the Bluetooth SIG.

2. *Cable replacement protocol (RFCOMM):* This one is constituted by Bluetooth SIG but it is based on the ETSI TS 07.10.

3. *Telephony protocol control specification (TCS BIN, AT-commands):* Also this one is constituted by Bluetooth SIG, but they are based on the ITU-T Recommendation Q.931.

4. *Adopted protocols (PPP, UDP/TCP/IP, WAP/WAE, OBEX, vCard, vCal, and IrMC).*

In addition, the specifications comprise a *host controller interface* (HCI), which provides a command interface to the baseband controller, *link manager controller* (LMC), and access to hardware status and control registers. In effect the Bluetooth core provides a common wireless system to various protocols, included those in this list and other ones freely implemented by vendors, which sometimes are interfaced with the core by a sort of convergence layer represented by RFCOMM.

6.3.3 Overview on Bluetooth Core Protocols

This section describes the different protocols that are the base of the Bluetooth standard, and in particular describes the baseband that provides functions and services nearest to those of the OSI MAC layer.

6.3.3.1 Bluetooth Radio Layer

The Bluetooth radio uses a *frequency hopping spread spectrum* (FHSS) system through 79 (Europe and United States) or 23 (France) subcarriers; the first subcarrier is located at 2.402 GHz (Europe and United States). The subcarriers are spaced at 1 MHz; radio specification defines lower and upper guard bands, which in Europe are 2 MHz–3.5 MHz. The channel is represented by a pseudorandom hopping sequence, which is unique for the piconet. It is determined by the Bluetooth device address of the master, which synchronizes the piconet with its clock. The nominal hop rate is 1,600 hops/second.

The modulation used in GFSK with a BT = 0.5.

6.3.3.2 Baseband Layer

The baseband layer provides a mapping of logical channels onto physical channels, which are defined over the time slot, each 625 µs in length and numbered according to the clock of the piconet master.

The system uses a TDD access method, which provides an alternate transmission between master and slave as shown in Figure 6.6.

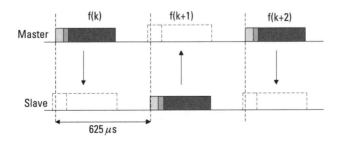

Figure 6.6 TDD and timing.

The RF hop frequency is fixed during the packet duration, even when packet duration is longer than a slot. The hop frequency for the next packet is counted as if a single packet per slot transmission occurred (see Figure 6.7).

The following services are offered by the baseband layer:

- Error correction with FEC and ARQ scheme (only for data packets);
- Data whitening;
- Transmission (Tx)/Reception (Rx)routines and timing;
- Flow control;
- Authentication and encryption;
- Management of audio transmission.

Physical Link

Master and slaves can be involved in the following different types of links:

- *Synchronous connection-oriented* (SCO) link, which is a point-to-point link between the master and a slave in a piconet;
- *Asynchronous connectionless link* (ACL), which is a point-to-multipoint link between the master and all of the slaves in a piconet.

We observe that a physical link is always started by the master or by a unit, after which it becomes the master of the piconet just formed.

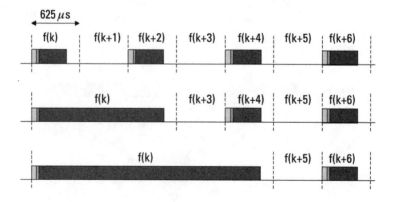

Figure 6.7 Multislot packets.

SCO Link

The SCO link reserves slots and can be considered as a sort of circuit-switched connection between the master and the slave. SCO link is used to support time-bounded information such as voice. The master can support up to three SCO links to the same slave or to different slaves; on the other hand, a slave can support up to three SCO links from the same master or two SCO links if the links originate from different masters. The master sends SCO packets at regular intervals, known as SCO interval T_{SCO} (counted in slots), to the slaves in the reserved master-to-slave slots. SCO packets are never retransmitted.

ACL Link

The ACL link does not reserve slots: It can be considered as a sort of packet-switched connection between the master and all active slaves in the piconet. ACL link is used for both asynchronous and isochronous services. Only a single ACL link can exist between a master and a slave.

A slave is permitted to return an ACL packet in the slave-to-master slot if and only if it has been addressed in the preceding master-to-slave slot. ACL packets that do not have an address of a specific slave are considered as broadcast packets and are read by all slaves. The ACL link provides packet retransmission to assure data integrity.

Packets

The general packet composition for a piconet channel is shown in Figure 6.8.

The packets are constituted only by access code, or access code-header or access code-header-payload.

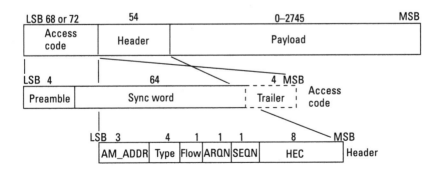

Figure 6.8   General Bluetooth frame format.

The access code is used for synchronization, dc offset compensation, identification, paging, and inquiry procedures. It always uses a preamble and sometimes uses a trailer for synchronization purposes.

The header contains the link control information and includes the destination address, the indication of one of the 16 types of packets, flow control information, the ACK indication, and the *header error check* (HEC).

Several payload formats are then defined, but two main fields are identified:

1. The (synchronous) voice field with fixed length where no payload header is present;

2. The (asynchronous) data field, which consists of three subfields: a payload header, a payload body, and possibly a CRC code.

The ACL packets only have the data field, while the SCO packets may contain only the voice field or have both. Finally, it is possible to distinguish a group of common packets for both ACL and SCO link, and a set of packets for each of them.

The bitstream processes, concerning the packet header and payload both at Tx and Rx, are described in Figure 6.9, where some blocks are optional, depending on the packet type. In particular, data whitening is performed to randomize the data from highly redundant patterns and to minimize dc bias in the packet.

Tx/Rx Routines

ACL and SCO links are managed differently by Rx/Tx routines: Both routines are based on some buffers for ACL and SCO that can be accessible for the Tx (Rx) in input (output) by the Bluetooth link manager and in output (input) by the packet composer, as shown in Figure 6.10.

Figure 6.10 shows only one pair of buffers for each entity, while in the master there is a separate Tx ACL buffer for each slave and a single Rx buffer is shared among all slaves. One or more Tx (or Rx) SCO buffers are present for each slave, depending on the different link established between them.

In the ACL buffers, only inserted ACL packets can be found. The same is true for the SCO buffers except for voice/data packets, which use ACL buffers for the voice and SCO buffers for the data.

Each Tx/Rx buffer consists of two *first input first output* (FIFO) registers: a current register and a next register.

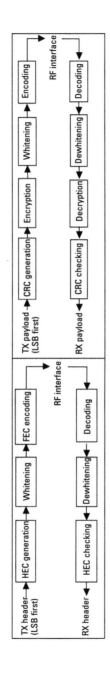

Figure 6.9 Header and payload bit processes.

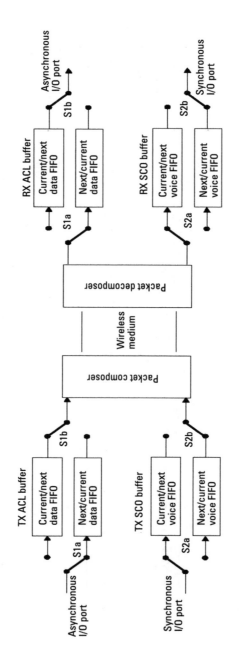

Figure 6.10 Functional diagram of Tx/Rx buffering.

Each switch shown in Figure 6.10 is controlled differently by each entity, but the switches at the input and the output of the FIFO registers can never be simultaneously connected to the same register. In particular, the switches that are linked with ACL buffers change position only when an ACK arrives; on the other hand, the switches linked with SCO buffers change according to the T_{SCO} interval. In this way, if the payload needs to be sent (received), the packet composer reads (writes to) the current register and forms (decomposes) the packet depending of its type.

The buffers have a finite number of positions, so that a *flow control* is necessary: it employs a STOP/GO indication to inform the source to freeze the transmission.

Master/Slave Synchronization

The clock of the master synchronizes the whole piconet. Following an indication from the master, all its slaves must adjust the system clock; this information is contained in the packets that it transmits. This mechanism ensures that a slave always follows the master behavior. It uses 28-bit counters where the LSB ticks in units of 312.5 μs.

In particular, a Bluetooth unit can use three clocks:

- Native clock (CLKN);

- Estimated clock (CLKE);

- Master clock (CLK).

CLK can be derived by the others as in the following relation: CLK = offset(CLKE) + CLKN where, in the case of master, offset is equal to zero.

Overview of States

A Bluetooth unit can stay in different states, on indication of LMP:

1. *Standby state:* this is the default state, where a unit is in a low-power mode and only the native clock is active.

2. *Connection state:* a unit enters in this state when it is involved in a communication. It provides several substates, the most important being:
   - *Active mode:* in this mode, the Bluetooth unit actively transmits information on the channel.

- *Sniff mode:* in this substate, the duty cycle of the slave's listen activity can be reduced according to the master.
- *Hold mode:* this is a substate of the connection mode where a slave can be placed by the master and where a Bluetooth transceiver does not transmit or receive information.
- *Park mode:* a slave is associated with the master, but it doesn't receive anything until it wakes up to listen to the beacons from the master and to resynchronize its clock offset. In this state, the slave is not involved in a connection.

When a slave wants to exit from the park mode, it listens to its specific beacon slots and tries to access to its specific access window, as shown in Figure 6.11.

In this case, a slave uses an AR_ADDR, which is not unique, so that collisions may occur.

This procedure can be avoided if the master sends an indication to unpark a parked slave with a dedicated LMP message.

In sniff mode, hold mode, and park mode, the transceivers are in low power consumption.

The connection state starts with an *access* command and is left through a *detach* or *reset* command.

Access Procedures

This procedure is used to establish new connections, and in particular it involves *inquiry* and *paging* procedures.

The inquiry procedure is executed when the destination device address is unknown to the source that sends a broadcast message. The inquiry message doesn't contain any information about the source, and it can be a *general inquiry access code* (GIAC) to inquire for any Bluetooth device or a *dedicated inquiry access code* (DIAC) to inquire only for a certain type of device. Such a

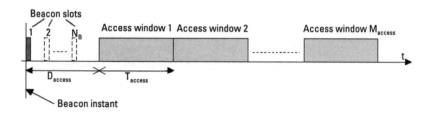

Figure 6.11 Definition of beacon channel and access window.

message is sent over a dedicated inquiry hopping sequence consisting of 32 frequencies.

A unit can respond only if it enters into an *inquiry scan* state. The sequence of inquiry procedure is shown in Figure 6.12.

On the other hand, the paging procedure is initialized by the master (source) that wants to connect with a slave (destination). It sends a paging message, which contains the destination address, into a specific paging hopping sequence consisting of 32 frequencies and tries to guess the right slave's phase.

The specific slave can respond only if it enters into a *paging scan* state. Then the master sends some synchronization information to the slave which, in turn, sends a response message.

At this moment the traffic packets synchronized with the master clock are transmitted, and a new piconet is formed if the master was still not communicating with other slaves.

The whole procedure is described in Figure 6.13.

Error Correction

Three types of error correction schemes are used:

- The first type is 1/3-rate FEC, employed for the header;

- A second type is 2/3-rate FEC, employing a (15,10) shortened Hamming code, which can be employed in the payload of ACL and SCO packets;

- The third type is ARQ scheme for the data.

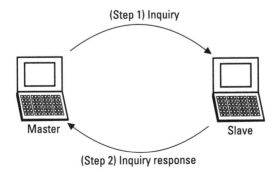

Figure 6.12   Inquiry procedure.

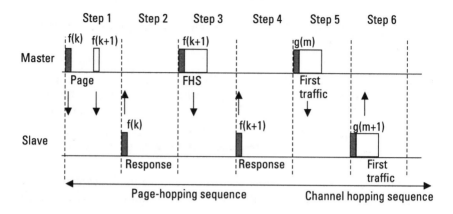

Figure 6.13 Paging procedure.

In particular, the employed ARQ scheme is a go-back algorithm, where packets are transmitted and retransmitted until the ACK (in piggybacking style) is returned by the destination or a time out is exceeded.

To determine if the packets are correctly received, a CRC code is added to the packet. The destination can also transmit a NACK indication whenever a corrupted packet is received.

To filter out the retransmissions in the destination, the SEQN bit is added in the header, and, if the packet arrives with an error, this bit is not changed. We emphasize that only data packets are protected with the ARQ mechanism, not voice payload or packet header.

With the particular case of isochronous traffic, where only a reduced delay is allowed, the algorithm permits a limited number of retransmissions after which old data is flushed (this determines the loss of remaining portions of an L2CAP message). The ARQ mechanism is not applied in the case of broadcast transmission, identified by the all-zero AM_ADDR, so that a master has to retransmit the same packet a fixed numbered of times before transmitting the next one.

Logical Channel

Bluetooth specifications define two control channels:

1. *Link control (LC) control channel:* this carries low-level control information like ARQ, flow control, and payload characterization. It is the only logical channel that is mapped onto the packet header.

2. *Link manager (LM) control channel:* this carries control information exchanged between the link managers of the master and the slave(s). It is mapped onto the payload.

Bluetooth system also defines three user channels:

1. *UA user channel (user asynchronous data):* this carries L2CAP user data, which can be segmented into one or more baseband packets. It is mapped onto the payload.
2. *UI user channel (user isochronous data):* this carries data that is supported by timing start packets property at higher levels. It is mapped onto the payload.
3. *US user channel (user synchronous data):* this is only carried over the SCO link and mapped onto the payload.

Bluetooth Addressing

It is possible to distinguish four types of addresses:

1. *Bluetooth device address (BD_ADDR):* this identifies each Bluetooth transceiver with a unique 48-bit address, which is derived from the IEEE 802 standard.
2. *Active member address (AM_ADDR):* this identifies each active slave (not the master) in the piconet with a 3-bit address. The all-zero address is reserved for broadcast messages. When a slave is disconnected or parked, it loses the AM_ADDR.
3. *Parked member address (PM_ADDR):* this identifies a slave in park mode and employs 3 bits. It is valid only as long as the slave is parked; when a slave is activated, it loses the PM_ADDR and takes an AM_ADDR.
4. *Access request address (AR_ADDR):* this employs a slave to identify the slave-to-master half slot in the access window to send access request messages. It is assigned by the master when a slave enters in the park mode.

Bluetooth Security

Bluetooth system mainly provides the management of security with the authentication and the encryption mechanisms. It also employs four different entities for maintaining security at the link layer: a public address, which is the 48-bit BD_ADDR; two secret user keys (one for the authentication and the other for the encryption, 128 bits and 8–128 bits, respectively); and a random number of 128 bits that is different for each new transaction.

Bluetooth Audio

Bluetooth system provides an air-interface to audio transmissions directly using the baseband layer and its ACL physical link. It uses either a 64 Kbps logarithmic PCM format (A-law or μ-law) or a *continuous variable slope delta* (CVSD) modulation, and the voice coding should have a quality equal to or better than the quality of 64-Kbps log PCM.

6.3.3.3   Link Manager Protocol (LMP) Layer

LMP messages are used for link setup, security, and control. LMP layer manages some services provided by the baseband layer, and it is transferred onto the ACL physical link instead of L2CAP. An LMP message is distinguished by a reserved value in the L_CH field of the payload header. LMP messages have higher priority than user data and are not delayed by the L2CAP traffic.

The services controlled by LMP are:

- Authentication/encryption;
- Synchronization;
- Management of unit states (hold mode, park mode, sniff mode);
- Power control;
- Quality of service;
- Paging scheme;
- Link supervision.

6.3.3.4   Logical Link Control and Adaptation Protocol (L2CAP) Layer

L2CAP provides connection-oriented and connectionless data services to the upper layers, and the essential requirements for L2CAP include simplicity and low overhead so that they can be implemented in devices with reduced computational resources. L2CAP has the following capabilities:

- *Protocol multiplexing:* this is to transmit multiple data flows onto the same physical link.
- *Segmentation and reassembly (SAR):* this is necessary to support protocols using packets larger than those supported by the baseband.

- *QoS:* L2CAP allows the establishment of connections with particular QoS and to monitor the resource in order to ensure that QoS contracts are honored.

- *Groups:* The L2CAP group abstraction permits implementations to efficiently map protocol groups into piconets. Without a group abstraction, higher-level protocols would be exposed to the baseband protocol and link manager functionality in order to manage groups efficiently.

The L2CAP specification uses only the ACL baseband link, and it employs those packets' formats that control their integrity (with CRC), because L2CAP does not ensure data integrity.

It is necessary to note that L2CAP protocol interferes with the packet composition performed in the baseband layer because it can change the flow bit in the packet header and the *logical channel* (L_CH) code (in the payload header) where there is an indication on the fragmentation status (an indication on start/continuation of L2CAP packet).

6.3.3.5 Service Discovery Protocol (SDP) Layer

SDP provides a means for applications to discover the available services and their characteristics, which depend on the RF proximity of the mobile device. The SDP layer provides a server-client interaction that gives to a client the ability to search for needed services that are based on some specific attributes and to require a server to discover a service on another device without consulting this one. The general architecture is shown in Figure 6.14.

6.4 PAN Paradigm

The person-centered networks continue to develop beyond the Bluetooth towards the dynamic network concept, which will allow easy communications with personal wearable devices and seamless movement within the existing network environment. The PAN approach is foreseen as a network paradigm, which attracts the interest of the researchers, and the industries grow toward the more advanced network solution, radio technologies, higher bit rates, enhanced terminals, new mobility patterns, and more sophisticated software support.

The PAN covers the area around the person recognized as a personal space. The general PAN network model is presented in Figure 6.15. It should

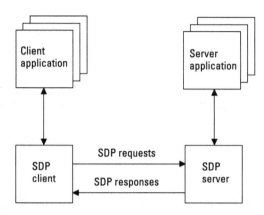

Figure 6.14 Overview of SDP protocol.

provide end-to-end connectivity, secure communications, and QoS guarantees to the users. The system should be able to support different applications and operating scenarios, and to comprise several devices.

6.5 Architecture Principles

The PAN is a network for you, for you and me, and for you and the outer world. To comprise these, it should develop layered architecture where different layers cover the specific types of connectivity (see Figures 6.16–6.18).

The connectivity is enabled through the incorporation of different networking functionality into the different devices. For the standalone PAN, the person should be able to address the devices within the personal space independent of the surrounding networks. For direct communication of two

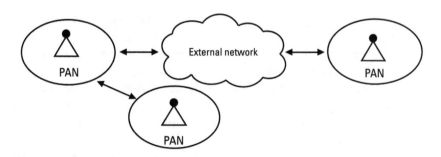

Figure 6.15 The PAN network concept.

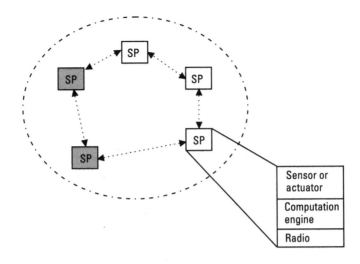

Figure 6.16 PAN is for you. A PAN constructs a personal sphere of smart peripherals (SPs).

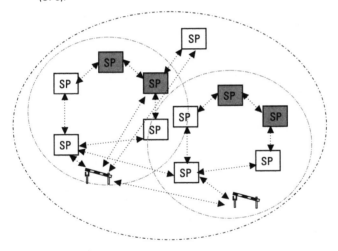

Figure 6.17 PAN is for you and me. When people and appliances meet, gatekeepers are needed. This is a dynamic distributed application platform.

persons (i.e., their PANs), the bridging functionality should be incorporated into each PAN. For communication through the external networks, a PAN should implement routing and/or gateway functionalities.

Layer-oriented scalable architecture should support the functionalities and protocols of the first three layers and should provide the capability to communicate with the external world through higher layer connectivity. It

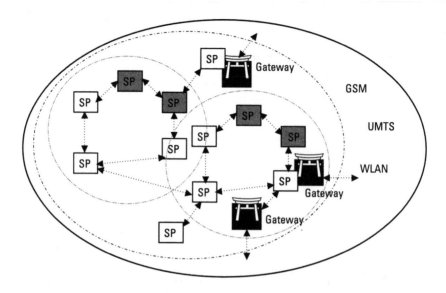

Figure 6.18 PAN is for you, me, and the outer world. Extending your reach requires a multimode gateway. This is distributed resource control with QoS.

should provide the appropriate middleware structures and consist of a well-defined protocol stack, with identified information transfer through appropriate interfaces.

The PAN should develop middleware structure capable of managing the system according to the access to the networks, resource discovery, support for scalability and reconfigurability, and QoS provisioning. It should also support the downloadable applications. The standardization of topology and architecture are still open issues within the PAN.

From the user point of view, PAN should offer plug-and-play connectivity. The network architecture should be seamless to the user.

Frequency planning and coexistence with the existing systems is important for designing novel PAN scenarios. The PAN-oriented applications will use the unlicensed frequency bands. For the higher data rates, the frequency band of 5 GHz, and perhaps later of 60 GHz, are foreseen (see Figure 6.19).

6.6   Interfaces

The PAN should support multiple interfaces. This system should introduce new types of interfaces between the person and his or her devices, two PANs,

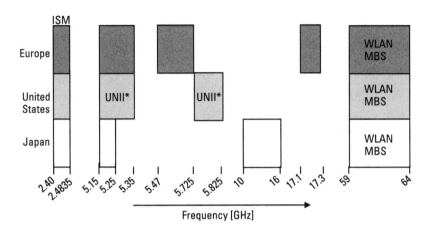

Figure 6.19 Frequency bands. (MBS: Mobile broadband system.)

and PAN and the outside world. This may rely on new technology solutions such as OFDM or DS-CDMA. The interface solutions will depend on appropriate data bit rate, network functionality, and complexity of the appropriate devices and applications.

6.7 Communication Via External Networks

The PAN mainly relies on wireless networking technologies. It can establish connections to the fixed infrastructures and external networks, such as WLAN, GSM, GPRS, UMTS, WLL, and *satellite* (S)-UMTS. The person can communicate via these networks and establish bilateral connections with another person/PAN or search on the network for desired information.

The continued growth of Internet traffic and information exchange influenced and shaped the PAN concepts. The future PANs will be the IP-based networks with support of the IPv6 protocols. They will introduce a new type of *domain mobility*. A variety of devices, applications, and services will raise different QoS demands towards availability, reliability, and security issues.

6.8 Ad Hoc Networking

To support the person's mobility, PAN implements the characteristics of the wireless mobile networks. The person, together with personal wearable devices, should be able to move freely and seamlessly and to be connected/disconnected to different network infrastructures in an ad hoc manner. The person should be able to enter the company space and to relay on WLAN, or to enter the car and to use its networking space without disrupting the ongoing connections. The personal space becomes a mobile domain capable of performing the ad hoc networking and multihoping. It should support mobile IP functionalities. The ad hoc features inherit dynamic resource allocation to cope with QoS requirements [13].

6.9 Security

Security is an extremely important issue within the PANs concept. It should cover the air interface, the software operations and operating systems, and the user's profiles. Different techniques such as encryption and clippering, *trusted third parties* (TTTs) mechanisms, and agent technology will be implemented. Security communications between the foreign PANs should be realized through the gatekeeping functions. The PAN security should offer:

- Full identity;
- Full anonymity;
- Data security;
- Integrity.

6.10 Main Applications and Possible Scenarios

The PAN searches for technical challenges and applications that will turn it into a system really serving the person and improving the quality of his or her personal and professional life. Many different operating scenarios can be foreseen, mainly concentrated around:

- Personal services;
- Business services;
- Entertainment.

The personal services include medical telemonitoring and control applications and smart homes. Medical professionals can constantly monitor the sick person wherever he or she moves. Within his or her home a person can move, addressing different devices and getting useful information from them.

The business scenario could consider building environmental monitoring, fleet management, and person searching. The content and temperature/pressure in the truck could be measured, as could the production line, the movement of the employees within the offices.

Entertainment scenarios could consider high-tech applications such as high-speed video on the trains, information retrieval through the personal devices, and other imaginable services. Applications will range from very low bit rate sensor and control data up to very high bit rate video streaming and bandwidth-demanding interactive games. The system should be able to support the variety and complexity of communication services and devices.

6.11 Possible Devices

Different service demands and application scenarios in PAN build up different approaches towards the terminal functionalities and capabilities. Some devices, such as simple personal sensors, must be very cheap and incorporate limited functionalities. Others should incorporate advanced networking and computational functionalities, which will make them more costly. Scalability regarding the following items is foreseen as crucial:

- Functionality and complexity;
- Price;
- Power consumption;
- Data rates;
- Trustworthiness;
- Supporting interfaces.

The most capable devices should incorporate multimode functionalities that will enable the access to multiple networks.

Some of the devices should be wearable or attached to the person (i.e., sensors); some could be stationary or associated temporarily to the personal space (i.e., environmental sensors, printers, and information desks). A set of possible devices for PAN applications is presented in Figure 6.20.

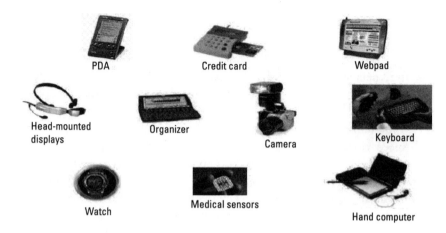

Figure 6.20 Possible devices for PAN applications.

6.12 PAN Challenges and Open Issues

Building the system with demanding system solutions for personal and network requirements stress many challenges to the scientific and industrial world.

PAN should build up applications on a dynamic distributed platform and provide distributed resource control with QoS. It should develop the low-power, low-cost radio and a variety of reconfigurable terminals. The PAN should blend into the living environment, allowing seamless network connectivity and secure communications.

To fulfill these goals, PAN should address important issues, such as:

- Low-power, low-cost radio integration;

- Definition of possible physical layers and access techniques;

- Ad hoc networking;

- Middleware architecture;

- Security (different security techniques, gatekeeping functionalities);

- Overall system concept;

- Human aspects.

Standardization activities regarding PAN are continuing within the IEEE 802.15 groups [10]. They will try to standardize the overall system concept and to point out the general requirements for PAN as a guidepost to the researchers and industry in bringing PAN into everyday life.

The consistent demand for the higher bit rates pushes the PAN concept towards the new border. The new network paradigm will present the *broadband PAN* (B-PAN) system, which will be able to support high data rates up to 400 Mbps and more demanding multimedia and broadband interactive applications.

6.13 B-PAN

The B-PAN is a future development of PAN towards the wideband adaptive novel techniques capable of broadband wireless communication (see Figure 6.21). It will support applications up to 1 Gbps and probably will operate over 5-GHz or 60-GHz frequency bands [3]. B-PAN will implement novel

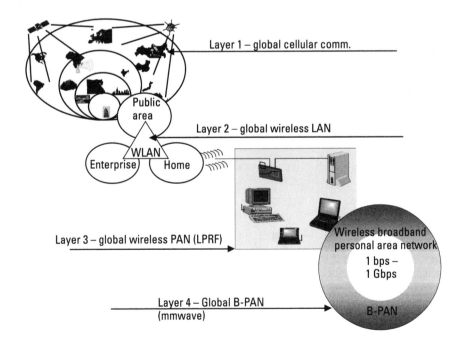

Figure 6.21 Four-layer wireless communication provides mobile services everywhere, and they complement each other. (LPRF: low power radio frequency.)

techniques such as *ultra wideband* (UWB), voice-over B-PAN, smart antenna, adaptive modulation, and coding, with extendable protocol functionalities. It should support performance QoS in adaptive and flexible manner. Different access methods and application interfaces will be defined, and the system will be supported with segmented intelligent multiaccess terminals capable of speech, messaging, and multimedia operations.

The B-PAN belongs to the wireless family that appears to be one of the most promising concepts, which opens tremendous possibilities for new applications. The technical differences between several previously mentioned wireless systems are presented in Table 6.2.

Table 6.2
Technical Differences and Applications

	UTRA	WLAN	Bluetooth	PAN	B-PAN
Data rates	Maximum 2 Mbps (384 Kbps)	5.1–54 Mbps	Maximum 721 Kbps	Maximum 10 Mbps	1 Gbps
Technology	TD-CDMA and W-CDMA	OFDM	DS or FH	OFDM	OFDM/DS-CDMA/SHF-CDMA
Cell radius	30m–20 km	50–300m	0.1–10m	To the distance a voice reaches	Similar to PAN
Mobility	High	Low	Very low	Very low	Very low
Standard availability	1999	2000	1999	2004	2012
Frequency band	2 GHz	5 GHz	2.4-GHz ISM band	5/10 GHz	60 GHz
Frequency license	Necessary	Not necessary	Not necessary	Not necessary	Not necessary
Application	Public environments (likely restricted use in places such as hospitals and airplanes)	Corporate environments (industrial applications); public hot spots (such as airports, exhibitions, convention centers)	Substitution for infrared communications; low cost networks for *small office home office* (SoHo) and residential applications	Personal peripheral device communications	Surrounding environment

6.14 WLANs Versus PANs

Regarding the rapid development of WLAN standards during recent years, as well as some of the target WLAN applications, a natural question arises: why is there research into PANs when there is already a well-traced line of progress for WLANs? WLANs can also afford wireless connectivity to the proximate portable computing devices, which is an initial drive for designing PANs. However, there are some important differences between WLANs and PANs.

WPAN technologies emphasize low cost and low power consumption, usually at the expense of range and peak speed. WLAN technologies emphasize higher peak speed and longer range at the expense of cost and power consumption. WLAN technologies emphasize higher peak speed and longer range at the expense of cost and power consumption. Typically, WLANs provide wireless links from portable laptops to a wired LAN via *access points*. To date, IEEE 802.11b has gained acceptance rapidly as a WLAN standard. It has a nominal open-space range of 100m and a peak over-the-air speed of 11 Mbps. Users can expect maximum available speeds of about 5.5 Mbps.

Although each technology is optimized for its target applications, no hard boundary separates how devices can use WPAN and WLAN technologies. In particular, as Figure 6.22 shows, both could serve as a data or voice

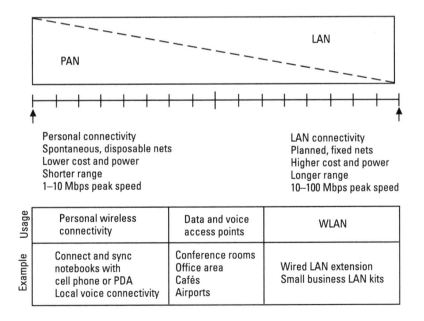

Figure 6.22 The complementary position of PANs and WLANs.

access medium to the Internet, with wireless WLAN technologies generally best suited for laptops and WPAN technologies best suited for cell phones and other small portable electronics.

A problematic topic about the Bluetooth PAN and IEEE 802.11b LAN is the coexistence issue [14, 15] because both operate in the unlicensed ISM band. When operated simultaneously in the same physical space, these two technologies degrade each other's performance.

Over the long run, researchers anticipate that WLANs will migrate to the 5-GHz unlicensed band, which may eliminate most coexistence issues. In particular, the companion standard IEEE 802.11a, designed for the 5-GHz band, will operate at peak over-the-air speeds up to 54 Mbps over distances up to 50m. Maximum data speeds available to users are projected to be between 24 and 35 Mbps.

6.15 Conclusions and Future Works

The PAN introduces the personal space concept into the communication world. It will develop toward the network extension within the personal world implementing a variety of new features in order to meet the rising service and network demands. The user surrounded by his or her personal smart space could move seamlessly and run various applications through the different network environments. Many issues are open to new solutions and ideas, which we may not even imagine at the moment. The B-PAN may be one of them.

References

[1]   Prasad, R., "Basic Concept of Personal Area Networks," *WWRF, Kickoff Meeting*, Munich, Germany, 2001.

[2]   Niemegeers, I. G., R. Prasad, and C. Bryce, "Personal Area Networks," *WWRF Second Meeting*, Helsinki, Finland, May 10–11, 2001.

[3]   Prasad, R., "60-GHz Systems and Applications," *2nd Annual Workshop on 60-GHz WLAN Systems and Technologies*, Kungsbacka, Sweden, May 15–16, 2001.

[4]   Prasad, R., and L. Gavrilovska, "Personal Area Networks," keynote speech, *Proc. EUROCON*, Vol. 1, Bratislava, Slovakia, July 2001, pp. III–VIII.

[5]   Prasad, R., and L. Gavrilovska, "Research Challenges for Wireless Personal Area Networks," keynote speech, *Proceedings of 3rd International Conference on Information, Communications and Signal Processing (ICICS)*, Singapore, October 2001.

[6] Zimmerman, T. G., "Personal Area Networks (PAN): Near-Field Intra-Body Communication," M.S. thesis, MIT Media Lab., Cambridge, MA, 1995.

[7] IBM PAN: http://www.almaden.ibm.com/cs/user/pan/pan.html.

[8] MIT Oxygen project: http://oxygen.lcs.mit.edu/.

[9] PicoRadio: http://www.gigascale.org/picoradio/.

[10] IrDA Standars: http://www.irda.com/.

[11] Bluetooth: http://www.bluetooth.com/.

[12] IEEE 802.15: http://grouper.ieee.org/groups/802/15.

[13] "Advances in Mobile Ad Hoc Networking," Special issue of *IEEE Personal Communications,* Vol. 8, No. 1, 2001.

[14] Howitt, I., "WLAN and WPAN Coexistence in UL Band," *IEEE Trans. on Vehicular Technology,* Vol. 50, No. 4, July 2001.

[15] Lansford, J., A. Stephens, and R. Nevo, "Wi-Fi (802.11b) and Bluetooth: Enabling Coexistence," *IEEE Network Magazine*, Vol. 15, No. 5, September/October 2001, pp. 20–27.

7

Paving the Way for 4G Systems

7.1 Introduction

The last chapter of the book introduces a solution, which was successfully designed and tested in the project denominated Wireless Internet Networks (WINE) financed by the European Union and belonging to the Fifth Framework Programme in the area of Information, Society, and Technology. It was aimed at two problems, as yet unsolved by the engineering community, which have been outlined in the previous chapters. The first is the provision of mechanisms enabling the interconnection of heterogeneous platforms exclusively through software developments. The second is the development of techniques that enable working with the TCP/IP stack over wireless links in the same way as over wired links. The solution of these two will undoubtedly make the path towards 4G easier.

The entity allowing the synergetic approach to these two objectives is called *wireless adaptation layer* (WAL), and, as has been mentioned, it has been tested successfully over infrastructures such as HIPERLAN, IEEE 802.11b, and Bluetooth. The following sections of this chapter are dedicated to presenting the most relevant aspects of the WAL architecture in the hope that it will serve as an example for the reader to formulate designs that really allow us to affirm that 4G has been reached.

7.2   Introduction to the WAL

The WAL can be defined as a PEP [1], designed for the protocols of the TCP/IP architecture when used over wireless networks. In Chapter 5, it was seen that the behavior of TCP and UDP protocols suffers a high degree of degradation over wireless technology. In this sense, the WAL attempts to compensate for these deficiencies, shielding the particularities of the wireless channel from the transport protocols. Thus, instead of modifying the existing protocols, which have a high penetration, the adaptation to be done by the WAL is situated between the IP layer and the underlying wireless infrastructure. In this way, and as the name indicates, the WAL adapts to the specific channel conditions, allowing the transport protocols to operate in their normal working mode.

The WAL is aware of the QoS required by IP (remember the ToS field in the IP header seen in Chapter 3), and adapts its function to be able to provide the appropriate service to satisfy it. If the IP layer does not implement any QoS mechanism, the WAL layer will use the traffic type identifier (protocol) to differentiate the different applications. In this way, the WAL complements the current policy of QoS in IP networks, *IntServ* [2] and *DiffServ* [3] (i.e., it does not duplicate its functionalities, but is aware of them).

The design of the WAL implies consideration of aspects relative to [4]:

- *Adaptation to the observed channel conditions.* The wireless channel conditions vary over time so the WAL will apply an adaptation scheme invoking the appropriate modules for each type of service and will modify its working parameters (for example, the correction capacity in the FEC module). For this, the information about the channel obtained by the WAL itself (e.g., SNR, BER, and throughput) is used, and this information is exchanged among the entities making it up.
- *Provision of the QoS required by IP.* The end-to-end QoS schemes used by IP (*IntServ* and *DiffServ*) are applied both in routers and in computers (the latest versions of the Linux kernel incorporate a traffic-control option). The WAL complements them in a local way, mapping the traffic requirements, to offer the appropriate service.

7.3   The WAL Architecture

Figure 7.1 shows the internal architecture of the WAL [4].

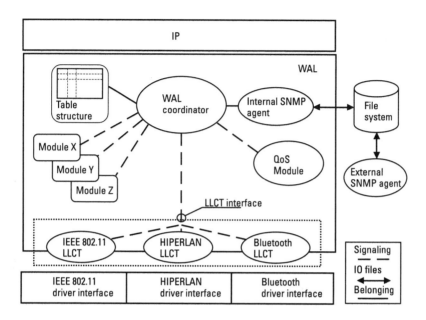

Figure 7.1 The internal WAL architecture.

As can be seen, the WAL is made up of different functional entities, among which the WAL coordinator should be highlighted. As its name indicates, it is the central element of the architecture. It manages the function of the rest of the modules, allowing them to access the structure tables. Furthermore, it will communicate with the coordinator of another remote WAL, with the aim of synchronizing the communication between the two.

Another essential WAL module is the QoS, which classifies the packets that arrive (it is the last model to be invoked) and passes them to the lower layers depending on their requirements.

With the aim of making the WAL function independent of the wireless infrastructure used, it is necessary to use an intermediate element, *logical link control translator* (LLCT). LLCT must carry out all of those functions necessary to guarantee the correct function of the system, on which the architecture used in the lower levels depends to some extent. Note that it is precisely this platform-dependent module that will allow the interoperability among heterogeneous infrastructures. In this sense, at the high level, the WAL provides an architecture for different wireless standards to integrate and communicate with each other. Additional differentiating features, such as the bridging function, can be added within this layer. With the introduction of

Bluetooth network encapsulation protocol in the Bluetooth PAN profile, which provides a clear interface between Bluetooth and IP, the WAL concept has become a very interesting part of the architecture.

A battery of modules is also used, each one with different characteristics, suitable for varied types of traffic. The WAL coordinator, taking into account the requirements of each packet it manages, will activate the corresponding entities, which in turn will adapt their working parameters depending on the channel conditions at any time.

The rest of the entities shown in Figure 7.1 are used to ease the implementation, as is the case of the file system and table structure, or to give additional functionality to the WAL, as is the case of the internal *Simple Network Management Protocol* (SNMP) agent module. These are not fundamental in order to understand its basic function.

7.4 WAL Signaling Services

Next, the communication protocol between WAL entities will be described in terms of the primitives interchanged. WAL entity parameters, the corresponding PDUs, their basic operation mode, and the procedures used in their function will also be detailed.

7.4.1 Some Definitions

The function of the protocol is based on two concepts: *class* and *association,* which are described next [4].

- *Class.* A class consists of the specific concatenation of a set of WAL modules with the aim of providing a service to the higher layers. It is determined from the ToS field of the IP header or the type of protocol (TCP or UDP). It is a general concept within the WAL, independent of the condition of the wireless channel. A series of general classes is defined, so that all of the WAL entities apply the modules corresponding to the packets belonging to them.

- *Association.* It has already been stated that each module adapts its function (modifying its parameters) depending on the conditions of the channel at any moment; therefore, it is necessary to introduce the concept of association, which is defined as the combination of the IP address of a mobile terminal (that determines the channel conditions) and a class.

7.4.1.1 The WAL Operation

The typical services provided by a link layer are of two types: connection oriented or nonconnection oriented. To satisfy some QoS requirements, the most appropriate choice is that of a connection-oriented service. In this sense, and taking into account that the connections based on the TCP/IP stack are identified by the combination of ports and origin and destination IP addresses, the overload caused by some applications would be unacceptable (for example, a visit to a Web page generates the opening of several TCP connections).

Bearing this in mind, it seems more suitable to implement the WAL with an association-oriented scheme. In this way, each time an IP datagram corresponding to a nonexistent association arrives at the WAL, a negotiation process is triggered that, if satisfactorily completed, generates a new association as a result. This association defines the suitable concatenation of the modules (class) and the working parameters to be applied from this moment to all the IP datagrams belonging to it.

7.4.1.2 WAL Format Header

All of the packets to be processed by the WAL will carry an associated header, whose format is shown in Figure 7.2.

The WAL header has a fixed size of 2 octets to facilitate the implementation in terms of byte handling. It identifies unequivocally all of the types of PDU through the PDU_Type field, of 6 bits, permitting up to 64 different types. The S/D bit allows the fast testing of whether a specific frame is a signalling (S/D = 1) or data (S/D = 0) frame. Finally, by the use of up to 7 bits in the Association_ID field, up to 128 different associations can be managed in a single AP.

7.4.1.3 Registration Procedure

From the WAL viewpoint, the first action to be considered is the registration process used in an AP to find out the WAL capacity of the mobile terminals entering in its coverage area. In this sense, it must be the LLCT that informs

WAL version	PDU_Type	S/D	Association_ID
2	6	1	7

Size (bits)

Figure 7.2 WAL header.

the WAL coordinator about the presence of a new MT, through a MAC-level registration.

Once the WAL coordinator is aware of the presence of a new MT in its coverage area, the WAL registration process is initiated. For this, the AP sends the WAL_CAPABILITY_REQUEST primitive, awaiting for a defined period of time a response from the MT, which must reply with a WAL_CAPABILITY_CONFIRM.

If the corresponding timer expires, the AP will resend the primitive and, if there is no response again, it supposes that the MT is not WAL capable. Both situations are described in Figure 7.3.

Figure 7.3 introduced the parameters that will have the primitives to be interchanged during the registration process, which are detailed in Figure 7.4.

Given that the design of the WAL attempts to maintain the compatibility both with IPv4 and IPv6, both primitives contemplate the possibility of using addresses of either format. When the AP launches the primitive, it still does not know which version the MT is using; therefore, it indicates both of its addresses (it may implement, simultaneously, both versions of IP). In that sense, it uses the address flag field to indicate what types of addresses are being managed (the bit 0 indicates the use of IPv4 and the 1 indicates the use of IPv6).

It is assumed that the MTs use only one version of IP, so there is only one address field in the reply primitive. The flag field indicates which of the two is being used (its size is one octet to facilitate the alignment and thus the implementation).

On the other hand, in order that the AP knows the two modules that the MT can use, a Class_List field of 2 octets is used, in which the classes that can be used are codified. For this, a class per bit is assigned, whose activation indicates the presence of a determined class, as is shown in Figure 7.5.

Once the WAL registration concludes, a signalling association must be established, which permits the WAL entities to interchange control messages. The AP, after receiving the WAL_CAPABILITY_CONFIRM, sends a primitive to the mobile terminal (SIGNALING_ASSOC_REQUEST). As the AP knows which modules are present in the MT, it accepts the configuration proposed for the association, responding with the primitive SIGNALING_ASSOC_CONFIRM. Moreover, in this way, the identifier of the association is provided by the AP, thus avoiding the risk of using repeated numbers.

The format of the primitives used in the creation of the signaling association is shown in Figure 7.6. The length of the fields used to determine the

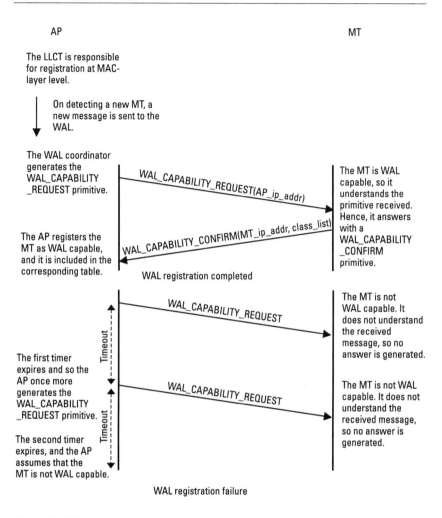

Figure 7.3 WAL registration procedure.

working parameters of the different modules depends on the configuration of each class, so their lengths cannot be determined.

Once the creation of the signalling association has concluded, both the MT and the access point will have an entry in their corresponding tables, with their characteristics (class and parameters of the modules), This is shown in Figure 7.7, which represents the interchange of primitives carried out in the previous process.

WAL_CAPABILITY_REQUEST

WAL header	@ flag	@IPv4 AP	@IPv6 AP

Size (bytes)	2	1	4	16

WAL_CAPABILITY_CONFIRM

WAL header	@ flag	@IPv4 MT @IPv6 MT	Class_ List

Size (bytes)	2	1	4 / 16	2

Figure 7.4 PDU interchange in the WAL registration procedure.

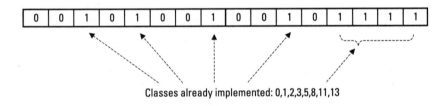

0	0	1	0	1	0	0	1	0	0	1	0	1	1	1	1

Classes already implemented: 0,1,2,3,5,8,11,13

Figure 7.5 Coding of the Class_List field.

SIGNALING_ASSOC_REQUEST

WAL header	Class_ID	Parameters Module_1		Parameters Module_N

Size (bytes)	2	1	?		?

SIGNALING_ASSOC_CONFIRM

WAL header

Size (bytes)	2

Figure 7.6 PDU interchange in the signaling association establishment.

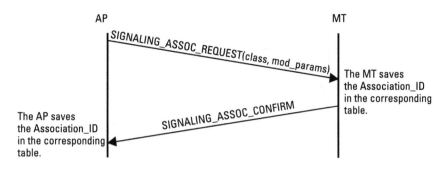

Figure 7.7 Signaling association establishment.

7.4.2 Association Establishment

This procedure is initiated each time that an IP datagram arriving at the WAL cannot be classified within the set of associations that the WAL is managing. Supposing that the MT initiates the process, it will send the primitive ASSOC_REQUEST, indicating the class and the working parameters of the modules making it up. The AP, after receiving the PDU, responds with the primitive ASSOC_RESPONSE, in which it offers certain working parameters that, in principle, could be different from those requested by the MT. If the configuration offered by the AP is satisfactory, the MT will confirm the creation of a new association, with the primitive ASSOC_CONFIRM. If, on the other hand, it refuses the configuration proposed, it would send the PDU ASSOC_REJ. To a certain extent, the interchange of messages carried out in the process (see Figure 7.8) is reminiscent of the three-way handshaking employed by TCP, which was explained in Chapter 3. In the event of the AP initiating the establishment of the association, the interchange of PDU would be similar.

The process described before is asymmetric, in the sense that it is always the AP that indicates the identifier for the association that is created, so overlaps are not produced. The AP knows all the IDs assigned for each MT, while the MT is unaware of those used by the AP to communicate with the rest of the mobile terminals within its coverage area.

As was commented previously, both the MT and the AP maintain all of the associations being managed in a table (see Table 7.1), which contains the class to which each one belongs, along with the working parameters of all the modules making it up. The AP must also relate each association to the corresponding MT, so it maintains an entry identifying the MTs (with their IP address, for example).

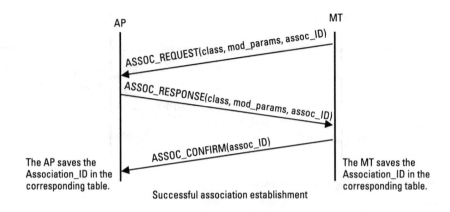

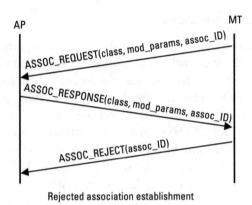

Figure 7.8 Association establishment procedure.

Table 7.1
Table Containing the Associations Already Established

Association_ID	Class_ID	MT_ID[1]	Parameters Mod_1
			Parameters_ Mod_2
			...
			Parameters_ Mod_N

1. Needed only in the AP.

Figure 7.9 shows the format of the PDUs interchanged during the process of creation of an association.

ASSOC_REQUEST

WAL header	Class_ID	Association _ID	Parameters Module_1		Parameters Module_N
Size (bytes) 2	1	1	?		?

ASSOC_CONFIRM and ASSOC_REJECT

WAL header	Association ID
Size (bytes) 2	1

Figure 7.9 PDU interchange in the association establishment procedure.

7.4.3 Data Interchange

When an IP datagram belonging to an already established association arrives at the WAL, the WAL coordinator generates a data PDU. Assuming that it does not require fragmentation, each module receives the packet and processes it, adding additional information if necessary (when control information must be transmitted to the corresponding module in the destination).

Figure 7.10 shows the format of the data PDU. The association it belongs to is indicated in the corresponding WAL header field. Besides, a field that is used to transmit fragmentation information (fragmentation information field) is added if it is necessary to segment the IP datagram. Its functionality is described next.

As has been commented, one of the principal aims of the WAL is to adapt to the radio channel conditions, in a transparent way for the higher layer protocols. In this sense, the size of PDU exported by the WAL must be fixed. However, it was seen in Chapter 5 that, depending on the state of the

Data PDU

WAL header	Fragmentation information	Module_1 data		Module_N data	IP datagram
Size (bytes) 2	1	?		?	?

Figure 7.10 Data PDU.

wireless link at any time, it could be interesting to modify the size of the PDU sent through the radio channel. Thus, it is necessary to implement some fragmentation mechanism. The octet reserved in the data PDUs (the only ones that can be fragmented, as the rest are not big enough) for this purpose has the format shown in Figure 7.11.

The entity managing the fragmentation for each association is responsible for the values used to number the WAL packets (WAL packet number) and its fragments (fragment number) not repeating. Finally, the bit M (more) is used to indicate that there are more fragments associated to the corresponding WAL packet number.

7.4.4 Reassociation Procedure

This procedure is invoked to modify the configuration of an association in the event of variation of the associated radio channel conditions—to adapt to the environmental conditions. Within the WAL statistics, the state of the wireless link must be collected, allowing the estimation of the current state (e.g., SNR). On the other hand, certain thresholds that mark the different working configurations for the associations of a specific class are defined. When the entity collecting the state of the channel detects that one of these thresholds has been crossed, it warns the WAL coordinator. This initiates the process of reassociation, similar to the association process described previously, as is shown in Figure 7.12.

After concluding the reassociation procedure, a new identifier is assigned to the resulting configuration. The primitives interchanged are, in this sense, identical to those used in the association-establishment process.

Finally, it should be remembered that the entity responsible for collecting channel statistics is the LLCT, given that the task depends strongly on the driver of the wireless interface used.

Fragmentation field

WAL packet number	Fragment number	Bit M
4	3	1

Size (bits)

Figure 7.11 Fragmentation information carried over a data primitive.

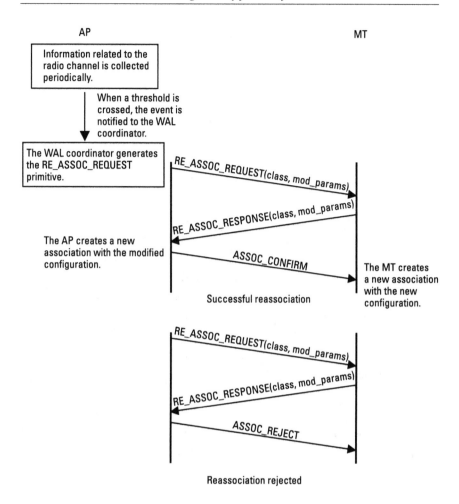

Figure 7.12 Reassociation procedure.

7.4.5 PDU List

To end this chapter, Table 7.2 shows the set of WAL primitives that have been used in the project WINE. In the same way, it should be noted that the design presented here is susceptible to future improvements, which are already underway in the framework of the project Protocols for Heterogeneous Multi-Hop Wireless IPv6 Networks (6HOP), the continuation of WINE. These basically include a WAL header length field and an error control field of this header.

Table 7.2
PDU List of the WAL

Type of PDU	Identifier	Fields	
		Name	**Size (Bytes)**
DATA	0x00	WAL header	2
		Fragmentation	1
		Data_Mod_1	n_1
		Data_Mod_N	n_N
		IP Datagram	?
WAL_CAPABILITY_REQUEST	001	WAL header	2
		Address flag	1
		IPv4 address[1]	4
		IPv6 address[1]	16
WAL_CAPABILITY_CONFIRM	002	WAL header	2
		Address flag	1
		IPv4 or IPv6 address[2]	4 or 16
SIGNALING_ASSOC_REQUEST	003	WAL header	2
		Class_ID	1
		Parameters_Mod_1	n_1
		Parameters_Mod_N	n_N
SIGNALING_ASSOC_CONFIRM	004	WAL header	2
ASSOC_REQUEST	005	WAL header	2
		Class_ID	1
		Association_ID	1
		Parameters_Mod_1	n_1
		Parameters Mod_N	n_N
ASSOC_RESPONSE	006	WAL_header	2
		Class_ID	1
		Association_ID	1
		Parameters_Mod_1	n_1
		Parameters_Mod_N	n_N
ASSOC_CONFIRM	0x07	WAL header	2
		Association_ID	1

Table 7.2 (continued)

Type of PDU	Identifier	Fields	
		Name	Size (Bytes)
ASSOC_REJECT	0x08	WAL header	2
		Association_ID	1
RE_ASSOC_REQUEST	009	WAL header	2
		Class_ID	1
		Association_ID	1
		Parameters_Mod_1	n_1
		Parameters_Mod_N	n_N
RE_ASSOC_RESPONSE	00A	WAL header	2
		Class_ID	1
		Association_ID	1
		Parameters_Mod_1	n_1
		Parameters_Mod_N	n_N
RE_ASSOC_CONFIRM	00B	WAL header	2
		Association_ID	1
RE_ASSOC_REJECT	00C	WAL header	2
		Association_ID	1

1. One address or both may appear.
2. The MT only implements one version of IP.

Finally, it should be highlighted that the set of modules that have been implemented up to now over IEEE 802.11b are the FEC, Snoop, fragmentation, and ARQ. The reader interested in studying in greater detail the WAL function and the performance of any of these modules should consult [5, 6].

References

[1]   Border, J., et al., "Performance Enhancing Proxies Intended to Mitigate Link-Related Degradations," RFC 3135, June 2001.

[2]   Braden, R., D. Clark, and S. Shenker, "Integrated Services in the Internet Architecture: An Overview," RFC 1633, June 1994.

[3]   Blake, S., et al., "An Architecture for Differentiated Services," RFC 2475, December 1998.

[4]   Mähönen, P., et al., "Platform-Independent IP Transmission over Wireless Networks: The WINE Approach," *IEEE Personal Communications*, Vol. 8, No. 6, 2001, pp. 32–40.

[5]   Muñoz, L., et al., "Optimizing Internet Flows over IEEE 802.11b Wireless Local Area Networks: A Performance-Enhancing Proxy Based on Forward Error Correction," *IEEE Comm. Mag.*, Vol. 39, No. 12, 2001, pp. 60–67.

[6]   Becchetti, L., et al., "Enhancing IP Service Provision over Heterogeneous Wireless Networks: A Path Toward 4G," *IEEE Comm. Mag.*, Vol. 39, No. 8, 2001, pp. 74–81.

About the Authors

Professor Ramjee Prasad received a B.Sc. in engineering from the Bihar Institute of Technology, Sindri, India, and an M. Sc. in engineering and a Ph.D. from Birla Institute of Technology (BIT), Ranchi, India, in 1968, 1970, and 1979, respectively. He joined BIT as a senior research fellow in 1970 and became an associate professor in 1980. While he was with BIT, he supervised a number of research projects in the area of microwave and plasma engineering.

Professor Prasad has also worked as a professor of telecommunications in the Department of Electrical Engineering at the University of Dar es Salaam (UDSM), Tanzania. At UDSM, he was responsible for the collaborative project Satellite Communications for Rural Zones with Eindhoven University of Technology, the Netherlands. He was also actively involved in the area of wireless personal and multimedia communications (WPMC) with the Telecommunications and Traffic Control Systems Group at the Delft University of Technology (DUT). Professor Prasad was the founding head and program director of the Center for Wireless and Personal Communications of the International Research Center for Telecommunications—Transmission and Radar. Since 1999, he has been the codirector of the Center for Person Kommunikation at Aalborg University, and also holds the chair of wireless information and multimedia communications. Professor Prasad was involved in the European ACTS project Future Radio Wideband Multiple Access Systems (FRAMES) as a DUT project leader. He is a project leader of several international, industrially funded projects. He has published over 300 technical papers, contributed to several books, and has authored, coauthored, and edited 11 books: *CDMA for Wireless Personal Communications, Universal Wireless Personal Communications, Wideband CDMA for Third-Generation Mobile Communications, OFDM for Wireless Multimedia Communications, Third Generation Mobile Communication Systems, WCDMA: Towards IP Mobility and Mobile Internet, Towards a Global 3G System: Advanced Mobile Communications in Europe, Volumes 1 & 2, IP/ATM Mobile Satellite Networks, Simulation and Software Radio for Mobile Communications and Wireless IP, and Wireless IP and Building the Mobile Internet*, all published by Artech House. His current research interests lie in wireless networks, packet communications, multiple-access protocols, advanced radio techniques, and multimedia communications.

Professor Prasad has served as a member of the advisory and program committees of several IEEE international conferences. He has also presented keynote speeches and delivered papers and tutorials on WPMC at various universities, technical institutions, and IEEE conferences. He was also a member of the European cooperation in the scientific and technical research (COST-231) project dealing with the evolution of land mobile radio (including personal) communications as an expert for the Netherlands, and he was a member of the COST-259 project. He was the founder and chairman of the IEEE Vehicular Technology/Communications Society Joint Chapter, Benelux Section, and is now the honorary chairman. In addition, Professor Prasad is the founder of the IEEE Symposium on Communications and Vehicular Technology (SCVT) in the Benelux, and he was the symposium chairman of SCVT'93.

In addition, Professor Prasad is the coordinating editor and editor-in-chief of the *Kluwer International Journal on Wireless Personal Communications* and a member of the editorial board of other international journals, including the *IEEE Communications Magazine* and *IEE Electronics Communication Engineering Journal*. He was the technical program chairman of the PIMRC'94 International Symposium held in The Hague, the Netherlands, from September 19–23, 1994, and of the Third Communication Theory Mini-Conference in Conjunction with GLOBECOM'94, held in San Francisco, California, from November 27–30, 1994. He was the conference chairman of the 50th IEEE Vehicular Technology Conference and the steering committee chairman of the second International Symposium WPMC, both held in Amsterdam, the Netherlands, from September 19–23, 1999. He was the general chairman of WPMC'01, which was held in Aalborg, Denmark, from September 9–12, 2001.

Professor Prasad is also the founding chairman of the European Center of Excellence in Telecommunications, known as HERMES. He is a fellow of IEE, a fellow of IETE, a senior member of IEEE, a member of the Netherlands Electronics and Radio Society (NERG), and a member of the Engineering Society in Denmark (IDA).

Dr. Luis Muñoz is an associate professor at the University of Cantabria. He received a telecommunications engineering degree from the E.T.S.E.T.B., Polytechnical University of Cataluña (UPC), Spain, in 1990, and a Ph.D. from the UPC in 1995. He has been working in the field of data transmission since 1990 in the areas of equalization, channel coding, medium access control, and data link control techniques. He later began to work in mobile networks designing and carrying out projects as terrestrial trunking radio (TETRA) for power utilities, security systems, and telecontrol with real-time needs. He has participated in ACTS projects, within the IV framework of the European Union research and development program, such as cellular access for broadband systems and interactive television (CABSINET). He is also involved in IST projects belonging to the V framework program, such as wireless Internet networks (WINE), power-aware communications for wireless optimized personal area networks (PACWOMAN), and protocols for heterogeneous multihop wireless IPv6 networks (6HOP). At present, Dr. Muñoz is also leading projects, in close cooperation with European operators and manufacturers, in the field of wireless IP.

He has more than 30 international publications in conferences and journals. His current research interest lies in performance enhancing proxies, 4G system, wireless local area networks, and wireless personal area networks.

Index

GSM Networks: Protocols, Terminology, and Implementation,
Gunnar Heine

GSM System Engineering, Asha Mehrotra

Handbook of Land-Mobile Radio System Coverage, Garry C. Hess

Handbook of Mobile Radio Networks, Sami Tabbane

High-Speed Wireless ATM and LANs, Benny Bing

Interference Analysis and Reduction for Wireless Systems,
Peter Stavroulakis

Introduction to 3G Mobile Communications, Second Edition,
Juha Korhonen

Introduction to GPS: The Global Positioning System,
Ahmed El-Rabbany

An Introduction to GSM, Siegmund M. Redl, Matthias K. Weber,
and Malcolm W. Oliphant

Introduction to Mobile Communications Engineering,
José M. Hernando and F. Pérez-Fontán

*Introduction to Radio Propagation for Fixed and Mobile
Communications,* John Doble

*Introduction to Wireless Local Loop, Second Edition:
Broadband and Narrowband Systems,* William Webb

IS-136 TDMA Technology, Economics, and Services,
Lawrence Harte, Adrian Smith, and Charles A. Jacobs

Mobile Data Communications Systems, Peter Wong and
David Britland

Mobile IP Technology for M-Business, Mark Norris

Mobile Satellite Communications, Shingo Ohmori, Hiromitsu
Wakana, and Seiichiro Kawase

*Mobile Telecommunications Standards: GSM, UMTS, TETRA, and
ERMES,* Rudi Bekkers

*Mobile Telecommunications: Standards, Regulation, and
Applications,* Rudi Bekkers and Jan Smits

Understanding Cellular Radio, William Webb

Understanding Digital PCS: The TDMA Standard,
  Cameron Kelly Coursey

Understanding GPS: Principles and Applications,
  Elliott D. Kaplan, editor

Understanding WAP: Wireless Applications, Devices, and Services,
  Marcel van der Heijden and Marcus Taylor, editors

Universal Wireless Personal Communications, Ramjee Prasad

WCDMA: Towards IP Mobility and Mobile Internet, Tero Ojanperä
  and Ramjee Prasad, editors

*Wireless Communications in Developing Countries: Cellular and
  Satellite Systems,* Rachael E. Schwartz

Wireless Intelligent Networking, Gerry Christensen,
  Paul G. Florack, and Robert Duncan

Wireless LAN Standards and Applications, Asunción Santamaría
  and Francisco J. López-Hernández, editors

Wireless Technician's Handbook, Andrew Miceli

For further information on these and other Artech House titles,
including previously considered out-of-print books now available
through our In-Print-Forever® (IPF®) program, contact:

Artech House	Artech House
685 Canton Street	46 Gillingham Street
Norwood, MA 02062	London SW1V 1AH UK
Phone: 781-769-9750	Phone: +44 (0)20 7596-8750
Fax: 781-769-6334	Fax: +44 (0)20 7630-0166
e-mail: artech@artechhouse.com	e-mail: artech-uk@artechhouse.com

Find us on the World Wide Web at:
www.artechhouse.com